Manufacturing Decision Support Systems

Manufacturing Systems Engineering Series

Series editor: Hamid R. Parsaei, Department of Industrial Engineering, University of Louisville, USA

The globalization of business and industry and the worldwide competitive economy are forcing business leaders in the manufacturing and service sectors to utilize fully the best equipment and techniques available. The objective is to have efficient control of the organizational structure in order to produce high quality products at lower prices within a shorter period of time.

Since the introduction of computers in the 1950s, Manufacturing Systems Engineering has experienced tremendous growth. The development of the discipline has helped industry to become more productive and to make more efficient use of resources. Manufacturing information systems, total quality management, facility layout, material handling, value engineering and cost analysis, safety, computer-integrated manufacturing, and production planning and shop floor control are just some of the areas in which manufacturing systems engineers have been traditionally involved in order to help improve understanding and awareness in the manufacturing and service sectors. The recent emphasis and concern about the environment and product recyclability and reusability have brought new perspectives and more challenges to this ever-growing engineering discipline.

The aim of the *Manufacturing Systems Engineering Series* is to provide an outlet for state-of-the-art topics in manufacturing systems engineering. This series is also intended to provide a scientific and practical basis for researchers, practitioners and students involved in manufacturing systems areas. Issues which are addressed in this series include, but are not limited to, the following:

- Production systems design and control
- Life cycle analysis
- Simulation in manufacturing
- Manufacturing cost estimating
- Industrial safety
- Fuzzy logic and neural networks in manufacturing
- CAD/CAM/CIM

We would welcome proposals to write material for this series from colleagues and industry leaders around the world. We hope that researchers both in academia and government, as well as private organizations and individual practitioners, will find this series informative and worthwhile.

1 **Manufacturing Decision Support Systems**
Edited by Hamid R. Parsaei, Sai Kolli and Thomas R. Hanley

2 **Integrated Product, Process and Enterprise Design**
Edited by Ben Wang

3 **Occupational Ergonomics**
Fariborz Tayyari and James L. Smith

Manufacturing Decision Support Systems

Edited by

Hamid R. Parsaei

Professor of Industrial Engineering
University of Louisville
USA

Sai Kolli

The Ultimate Software Consultants
Lombard
USA

Thomas R. Hanley

Professor of Chemical Engineering
University of Louisville
USA

CHAPMAN & HALL

London · Weinheim · New York · Tokyo · Melbourne · Madras

Published by Chapman & Hall, 2–6 Boundary Row, London SE1 8HN, UK

Chapman & Hall, 2–6 Boundary Row, London SE1 8HN, UK

Chapman & Hall GmbH, Pappelallee 3, 69469 Weinheim, Germany

Chapman & Hall USA, 115 Fifth Avenue, New York, NY 10003, USA

Chapman & Hall Japan, ITP-Japan, Kyowa Building, 3F, 2–2–1 Hirakawacho, Chiyoda-ku, Tokyo 102, Japan

Chapman & Hall Australia, 102 Dodds Street, South Melbourne, Victoria 3205, Australia

Chapman & Hall India, R. Seshadri, 32 Second Main Road, CIT East, Madras 600 035, India

First edition 1997

Typeset in 10/12 pt Palatino by Words & Graphics Ltd., Anstey, UK
Printed in Great Britain by T.J. Press Ltd., Padstow, Cornwall, UK

ISBN 0 412 57040 8

A catalogue record for this book is available from the British Library

Library of Congress Catalog Card Number: 96–84880

∞ Printed on permanent acid-free text paper, manufactured in accordance with ANSI/NISO Z39.48-1992 and ANSI/NISO Z39.48-1984 (Permanence of Paper).

Contents

Contributors

Adedeji B. Badiru
Expert Systems Laboratory,
School of Industrial Engineering,
University of Oklahoma,
Norman,
Oklahoma 73019,
USA

Prashant Banerjee
Department of Mechanical Engineering (M/C 251),
2039 Engineering Research Facility,
University of Illinois,
842 W. Taylor,
Chicago,
Illinois 60607,
USA

Qais Barakat,
Department of Mechanical Engineering (M/C 251),
2039 Engineering Research Facility,
University of Illinois,
842 W. Taylor,
Chicago,
Illinois 60607,
USA

S. Daniel Bird
Department of Administration,
Albers School of Business and Economics,
Broadway and Madison,
Seattle University,
Seattle,
Washington 98122–4460,
USA

Tobias Canz
Department of Systems and Industrial Engineering,
University of Arizona,
Engineering Building #20,
Tucson,
Arizona 85721,
USA

Denis R. Cormier
Department of Industrial Engineering,
North Carolina State University,
Raleigh,
North Carolina 27695–7906,
USA

James F. Cox III
Department of Management,
The University of Georgia,
419 Brooks Hall,
Athens,
Georgia 30602,
USA

Ali F. Emdad
Department of Information Sciences and Systems,
Morgan State University,
Cold Spring Lane and Hillen Road,
Baltimore,
Maryland 21239,
USA

Sunderesh S. Heragu
Department of Decision Sciences and Engineering Systems,
Rensselaer Polytechnic Institute,
Troy,
New York 12180–3590,
USA

Sanjay S. Jagdale
Department of Systems and Industrial Engineering,
University of Arizona,
Engineering Building # 20,
Tucson,
Arizona 85721,
USA

Sundaresan Jayaraman
School of Textile and Fiber Engineering,
Georgia Institute of Technology,
Atlanta,
Georgia 30332,
USA

Raj Jayashankar
Department of Industrial Engineering,
Northern Illinois University,
DeKalb,
Illinois 60115–2854,
USA

Sivakumar R. Kakuturi
Burrows Paper Corporation,
501 West Main Street,
Little Falls,
New York 13363,
USA

Suleiman K. Kassicieh
Anderson School of Management,
University of New Mexico,
Albuquerque,
New Mexico 87131,
USA

Murali Krishnamurthi
Department of Industrial Engineering,
Northern Illinois University,
DeKalb,
Illinois 60115–2854,
USA

Ratnesh Kumar
Department of Electrical Engineering,
University of Kentucky,
Lexington,
Kentucky 40506–0046,
USA

Reuven R. Levary
Department of Management and Decision Sciences,
School of Business and Administration,
3674 Lindell Boulevard,
St. Louis,
Missouri 63108,
USA

Haomin Li
Department of Electrical Engineering,
University of Kentucky,
Lexington,
Kentucky 40506–0046,
USA

Archie Lockamy III
School of Business and Industry,
Florida A&M University,
Tallahassee,
Florida 32307,
USA

Sambasivan Narayanan
Yellow Integrated Logistics,
Yellow Freight Systems,
10990 Roe Avenue,
Overland Park,
Kansas 66211–121,
USA

Peter J. O'Grady
Department of Industrial and Management Engineering,
University of Iowa,
Iowa City,
Iowa 52242,
USA

John W. Peters
Morris-Knudsen, Inc.,
101 Convention Center Road,
Las Vegas,
Nevada 89109,
USA

Don T. Phillips
Department of Industrial Engineering,
Texas A&M University,
College Station,
Texas 77843,
USA

H.V. Ravinder
Anderson School of Management,
University of New Mexico,
Albuquerque,
New Mexico 87131,
USA

Fahimeh Rezayat
School of Business Administration,
California State University – Dominguez Hills,
Carson,
California 90747,
USA

Steven A. Yourstone
Anderson School of Management,
University of New Mexico,
Albuquerque,
New Mexico 87131,
USA

Jigish S. Zaveri
Department of Information Sciences and Systems,
Morgan State University,
Cold Spring Lane and Hillen Road,
Baltimore,
Maryland 21239,
USA

Preface

During the last two decades, a tremendous growth in the popularity and applications of computers in manufacturing has occurred. Computer-aided design, computer-aided manufacturing, flexible manufacturing systems, group technology and many others are considered by many manufacturing executives as the most promising technologies and philosophies that, if successfully implemented, can reduce costs and enable the US manufacturing companies to become more competitive in the global market. In the computer-integrated manufacturing environment, the decision processes are often more involved. The decision makers are frequently required to have access to a vast amount of data to support and analyze their complex decision problems at strategic and tactical levels.

Decision support systems are often referred to as computer-based information technologies that allow the decision makers to interactively communicate and solve the decision problems. *Manufacturing Decision Support Systems* is intended to report the latest developments and address the central issues in this area. This volume consists of 14 refereed chapters, written by leading researchers from academia and industry.

In the first chapter, Krishnamurthi *et al.* present a prototype simulation-based cost analysis system that uses activity-based costing techniques to estimate the base manufacturing costs of products. This system uses the results generated by the simulation to analyze and generate an estimate for the base manufacturing costs of the products manufactured. In the second chapter, Kassicieh *et al.* describe systems used for computer-integrated manufacturing justification providing information about the elements of a proposed decision support system (DSS). The third chapter, by Lockamy and Cox, demonstrates how integrated performance measurement systems can be used as a manufacturing decision support system (MDSS) to support strategic planning processes in manufacturing environments. In Chapter 4, Bird addresses the application of support systems technology to the processes of quality management for the construction of intelligent quality management decision aids. In the fifth chapter, Narayanan and Jayaraman present the details of the design and implementation of a knowledge-based system that can assist decision makers in evaluating apparel manufacturing enterprises. Chapter 6, by

Peters and Badiru, presents a heuristic database structure for diagnostic expert systems in manufacturing. The approach is based on the heuristic matrix methodology developed by Peters. The database structure facilitates the handling of data uncertainty in diagnostic expert systems. The proposed methodology permits the utilization of data that is resident or could be entered, retained and retrieved from a company's management information systems database. The seventh chapter, by Banerjee and Barakat, identifies an object-oriented organization of network flow problem solving knowledge for manufacturing decision support systems. In Chapter 8, Jagdale and Canz describe the application of machine-learning techniques to a decision-making process in manufacturing automation. The objective of this chapter is to show that machine-learning techniques can be applied to represent machine-specific performance behavior of machine tools for a 'knowledge base' in intelligent machining. Heragu and Kakuturi, in Chapter 9, present an interactive program for machine grouping and layout. Their three-stage approach discussed in this chapter has been used to solve a number of test problems and an industrial problem. Chapter 10, by Zaveri and Emdad, introduces a genetics-based machine learning system that has been designed to search for a useful sequencing/scheduling heuristic. In Chapter 11, Li and Kumar develop a technique based on integer programming to determine both an optimal assignment of pickup locations as well as an optimal sequence of pickup and placement of components for printed circuit board (PCB) assembly. Rezayat, in Chapter 12, introduces an application of neural networks for the integration of statistical process control and engineering process control. A new approach to design improvement based on new algorithms that extend the conventional genetic algorithm to handle mixtures of algebraic, conditional, and database constraints with quantitative, symbolic, and qualitative variables is presented by Cormier and O'Grady in Chapter 13. Levary, in the last chapter of this volume, describes a conceptualized information system that integrates the programmable design of a product and its manufacture with both planning activities and manufacturing processes.

Hamid R. Parsaei
Sai Kolli
Thomas R. Hanley
Louisville, Kentucky, USA
November 1996

Acknowledgements

We are indebted to our authors and reviewers for their outstanding contributions and assistance in preparing this volume. We would also like to thank Drs Herman R. Leep and Mickey R. Wilhelm of the University of Louisville for their invaluable support and advice. A special word of thanks is due to Dr Dongke An and Mr Yung-Nien Yang for providing exceptional help to make this endeavor possible. Finally, we would like to express our deepest gratitude to Mr Mark Hammond of Chapman & Hall for giving us the opportunity to initiate this project.

CHAPTER ONE

A generalized cost analysis system for manufacturing simulation

Murali Krishnamurthi, Raj Jayashankar and Don T. Phillips

1.1 INTRODUCTION

Simulation is widely used as an effective decision support tool in a variety of applications ranging from flight scheduling (Haeme, Huttinger and Shore, 1988) to knowledge-based simulations for production scheduling (Muller, Jackman and Fitzwater, 1990). This wide applicability and popularity can be attributed to the commercially available sophisticated simulation languages and packages and their flexibility in being able to interface and integrate with existing decision-making tools and techniques.

Simulation is one of the most popular techniques currently being used to model manufacturing systems and evaluate the impact of changing various system parameters. It is obvious that increasing, for example, the number of units of a resource will generally result in a better throughput; but how much this change will cost with respect to the cost of reduced throughput due to the bottleneck is often difficult to determine. This is because the introduction of new resources will increase the cost of production. Therefore, it becomes important for decision makers to know the effect of these changes not only on use of resources and other production parameters but also on the final product allocated cost.

The economic consequences of implementing the conclusions derived from manufacturing simulations have been traditionally explored using financial simulations which add cost estimates to simulation results and evaluate their impact on the financial bottom line (O'Loughlin, 1990). Even though this approach addresses some of the needs in this area, it lacks the techniques and features to perform a comprehensive cost analysis of manufacturing systems. The most significant point in favor of including cost parameters in manufacturing simulation according to O'Loughlin is that 'when models where cost information is used for planning and development of the production line, was compared to the actual cost of the system after development, they proved to be within 5% to 15% of the modeling results. This allowed the developers to accurately predict and control the budget through all development phases of the system.'

Product costing has been one of the more difficult tasks in manufacturing accounting. Due to the sheer number and complexity of manufacturing activities, the task of accurately allocating product costs has inherently been cumbersome. Accounting costs do not reflect reality and are hence a hindrance to decision making. This is because financial product costs, which are compiled by accountants, generally distribute overhead costs in proportion to direct labor. Consequently low volume complex products are charged less than their share of the indirect costs (Beaujon and Singhal, 1990). Kaplan (1989) suggests a dynamic system of cost allocation rather than allocating costs broadly across cost categories. This technique will directly attribute costs to the value of the activities. One such technique is activity-based costing (ABC) which is increasingly being used as an alternative to traditional cost accounting techniques.

Activity-based costing involves the identification of activities and the cost incurred due to their performance. The cost of an activity is determined by using activity measures such as labor hours and production quantity. There are two ways to accurately estimate product costs. The first is to supplement traditional costing techniques with concepts such as activity-based costing. The second is to augment manufacturing simulations with appropriate cost collection techniques. Discrete event simulation languages are ideally suited for the latter since simulation time is advanced from one event to the next, and cost collection routines can easily access the duration of activities that occur between these events. This compatibility between the two has provided the added impetus to explore the possibility of developing a generalized cost analysis system for manufacturing simulation.

In this work, the feasibility of developing a generalized simulation-based cost analysis system that aids manufacturing decision making is explored and a framework is proposed for accomplishing this objective. Based on this framework, a prototype cost analysis system that estimates base manufacturing costs of products using the information generated by simulation models of the manufacturing systems has been designed, developed and implemented. The prototype system consists of a cost analysis module built around a commercial simulation language without requiring changes to the source code of the language or its existing structure. The costing system is transparent to the user, so that the user is required to do only a minimum amount of additional programming in the simulation language, thus providing an alternative to modeling the cost functions in the simulation language itself. The feasibility of the cost analysis system has been validated by estimating the product costs for a hypothetical manufacturing scenario using the cost analysis system and comparing it to product costs estimated for the same scenario by computing the costs within the simulation model itself.

In the prototype system, activity-based costing techniques are modeled using the C++ programming language and the production activities are tracked using models developed in the simulation language, SIMAN (Pegden, Shannon and Sadowski, 1990). The data thus generated by the simulation are used by the cost module to estimate product costs. The prototype system does not report all manufacturing costs, but only those costs accumulated by a product as it is processed through a manufacturing system that has been modeled using the simulation language.

1.2 REVIEW OF PAST WORK

While traditional simulations have dealt with evaluating performance measures, such as resource usage and throughputs, very little attention has been paid to costs. The idea of including a cost component to manufacturing simulation languages was first explored by Pritsker in the development of Q-GERT (Whitehouse, 1973). Few simulation languages or packages have specific provisions for including costs and investment parameters and those that do rarely represent proper accounting practices (McLanahan and Ketcham, 1990). The following paragraphs discuss past work germane to this research.

The IBIS simulation package developed at the University of Massachusetts, Amherst, demonstrates cost modeling abilities but does not have the necessary features to model complex production systems (Ketcham, Shannon and Hogg, 1990). The important feature of this simulated accounting system is that information flow is separated from production flow. The other point of interest is that information is recorded by entities through their attributes and this results in increasing the complexity of the simulation program and reducing its speed.

The concept of integrating production and finance was explored by Naylor (Christy and Kleindorfer, 1990) in the development of SIMPLAN. SIMPLAN is more of a financial planning package than a simulation language. Hence the modeling of complex production systems was beyond its scope. Though it used time-advance techniques to update variables, it did not have the sophisticated features of current commercial simulation packages to model realistic manufacturing systems.

Monahan and Smunt (Christy and Kleindorfer, 1990) describe a decision support system that evaluates the shift from batch processing to automated flexible processing, incorporating cost functions and some simulation components that determine the utilization of capacity. The model was developed in the XCELL simulation language and the cost parameters were built into the model. This approach addressed a specific problem, but is not capable of accurately representing the complexities of production processes.

The review of past attempts at integrating cost analysis into manufacturing simulations reiterates the fact that there is scope for additional research in this area. With the advances in simulation languages and their ability to collect detailed statistics and represent them in convenient formats, and with the simultaneous advances in cost accounting, the need to explore better ways to integrate the two seems imperative. An obvious solution would be to develop a simulation language which has the means of effectively modeling cost parameters. An immediate solution would be to develop a cost analysis system which can use the statistics generated by commercial simulation packages. The feasibility of developing such a system to support manufacturing decision making is explored in this effort.

1.3 DESIGN OF THE COST ANALYSIS SYSTEM

There are two closely connected problems in the conceptualization and design of a cost analysis system. One has to do with accurately allocating the costs of the manufacturing system to the products manufactured in it. The other has to do with accurately modeling the production environment so that it resembles the real-world system. The flexibility and the complicated nature of manufacturing systems render it difficult to address both issues simultaneously. A framework that possesses manufacturing modeling capabilities generally has limited cost-modeling abilities, and on the other hand, a financial-modeling framework may not do justice to the needs of modeling complex manufacturing systems. Both the cost-modeling and manufacturing aspects can be effectively integrated by capturing the manufacturing details in a general purpose simulation language and building the cost-modeling features around it (Christy and Kleindorfer, 1990). This research is primarily directed towards using a commercial simulation language to model the manufacturing system, and interfacing a cost analysis system which will use the output generated by the simulation model to estimate production costs.

The design and development of a framework for interfacing a cost analysis system with manufacturing simulation must address the following issues.

1. The algorithms necessary to estimate product costs using activity-based costing should be developed and it should be decided whether costs should be estimated off-line or on-line.
2. The system should be capable of storing and accessing a sizable volume

of data that is required to describe the production process. The system should also be able to access the statistics generated by the simulation program to be used during the estimation of product costs.
3. The system should be user friendly and require a minimal amount of additional programming by the user.
4. The constructs for modeling and collecting the cost information should be identified in the chosen simulation language.
5. The overall features of the system for controlling the interfaces and the storage and retrieval of data must be specified.

The following sections address these issues in detail before discussing the conceptualization of a comprehensive framework for interfacing a cost analysis system with manufacturing simulation.

1.3.1 On-line versus off-line cost modeling

One of the crucial design issues in the development of a cost analysis system is where to collect the cost information. Information pertaining to product costs can be collected on-line (that is, inside the simulation), or can be computed off-line (that is, outside the simulation) using the outputs generated from simulation. Moore (1990) argues that the benefits of modeling cost parameters outside of the simulation model far outweigh the advantages of doing it on-line. In particular, he lists the following reasons:

1. Calculations of cumulative costs during the simulation take up precious space in the form of additional attributes, and increase the number of computations in the form of cost calculations. This increases the simulation run time by as much as 20–50 percent.
2. The information is non-reusable. That is, once the simulation has been completed evaluating the effects of changes in the cost factors requires another complete simulation run. If the cost parameters are outside of the simulation, calculations that would otherwise require many hours would require only a few minutes to re-calculate from simulation statistics generated already.
3. Finally, if an adequate cost model can be built independent of the actual simulation, but fed by the statistics from the simulation, then the same cost model can be used in conjunction with a real-time data collection system. This would make it a useful tool for controlling the daily operations at the factory floor level.

These reasons, along with the fact that the cost analysis system should be a generalized one, have favored the design of an off-line cost analysis system.

1.3.2 Formulation of the cost-modeling functions

Base production cost of a product is the sum of the cost of raw materials, operators, use of resources (e.g. machines) scrap and consumables (e.g. coolants and fluids). To estimate the entire cost of manufacturing a product, the cost of machine and operator idle time, and factory overhead costs should also be allocated to the product. Some of the costs such as machine downtime and operator idle-time costs are also difficult to allocate. It is perhaps too simple, and perhaps even counterproductive, to artificially allocate these costs directly to a product. It may also be a good idea to identify these costs and leave them unallocated (Moore, 1990).

Activity-based costing involves the identification of various activities performed in an organization and the determination of their costs. Activities are recorded as actions and their cost is determined in measures, such as labor hours and processing times. The major costs associated with a production process are the cost of using machines, labor and consumables used in the production process. The cost of using a machine is driven by the number of time units a product uses it. From the estimates of the hourly cost of using a machine, the cost accumulated by a product as it uses the machine can be calculated. The labor use cost would be the cost of the operator needed for processing the product, during each manufacturing step. The cost of consumables used in the production process would have to be accumulated for each processing step. The cost drivers that drive these costs are discussed in the following subsections.

(a) Machine-time-based cost

This is the hourly cost of using a machine and includes fixed costs, such as purchase, installation and maintenance apportioned over the life of the machine. The fixed cost of using a machine is

$$H_i = \frac{[(P_i + I_i)/L_i] + M_i + A_i}{W} \tag{1.1}$$

where H_i is the hourly cost of using machine i; P_i is the purchase cost of machine i; I_i is the installation cost of machine i; L_i is the life in years of machine i; M_i is the annual maintenance cost of machine i; W is the number of working hours of machine i per year.

The user is required to input the parameters required for the estimation of the hourly cost of using the machine. This is used to estimate the hourly cost of using the machine by each of the products.

(b) Machine usage cost

This is the cost of processing all the steps of a product type on the various machines as represented by the process flow. This cost is computed using the machine usage time and the fixed cost of using each machine:

$$M_j = \sum_{k=1, i\varepsilon A, j\varepsilon B}^{Y} \frac{(s_{i,j,k} + l_{i,j,k} + p_{i,j,k} + u_{i,j,k} + qc_{i,j,k})Q_{i,j,k}}{Q_{j,k}} H_i \tag{1.2}$$

where M_j is the machine usage cost for all steps of product lot type j; s, l, p, u, qc are the set-up, load, processing, unload, and quality control times, respectively; subscripts i, j, k refer to an activity performed at machine i, for step k of product type j; Y is the total number of processing steps for product j; A is the set of machines and B is the set of products; Qi, j, k is the number of lots of job type j that used machine i for processing step k in the process flow; Q_j, k is the total number of good lots of product j processed in step k.

The sum of the set-up, load, processing, unload and quality control time is averaged over the number of good lots of product type j. This means that the costs accumulated by lots that scrapped at each processing step are apportioned to the lots that pass inspection and proceed to the next step. The time that each product type used a particular machine should be maintained by variable along with the total number of lots processed and the good lots that pass inspection. This average is a statistic that must be generated by the simulation, and accessed by the system.

(c) Labor usage cost

This is the cost of all the operators used for all processing steps of a product lot. This cost is computed using the hourly cost of the operator. The formulation of this cost is

$$O_j = \sum_{k=1, i\varepsilon C, j\varepsilon D}^{Y} \frac{(s_{i,j,k} + l_{i,j,k} + m_{i,j,k} + i_{i,j,k} + t_{i,j,k})Q_{i,j,k}}{Q_{j,k}} OC_i \tag{1.3}$$

where the subscripts i, j, k refer to an activity performed by operator i for step k of product type j; Y is the total number of processing steps for product type j; O_j is the labor cost for product type j; s, l, m, u, i, t stand for set-up, load, monitor, unload, inspection and transport time; OC_i is the hourly wage of operator i; $Q_{i,j,k}$ is the number of lots of product type j monitored by operator i for step k and $Q_{j,k}$ is the total number of good lots of type j processed at step k.

As in the previous case the entire time that a product used a particular labor resource is averaged over the total number of good lots of the product that passes inspection at that stage. This is another statistic the simulation must capture and report.

(d) Consumable usage cost

A consumable is any material that is used in the production process. A consumable could be a coolant used in a machining operation, or a chemical used in a production process. The formulation used in the estimation of the consumable usage cost is

$$C_j = \sum_{k=1, i \varepsilon E, j \varepsilon F}^{Y} \frac{(n_{i,j,k} c_i) Q_{i,j,k}}{Q_{j,k}} \tag{1.4}$$

where, C_j is the consumable usage cost of product type j; $n_{i,j,k}$ is the number of consumable i required per lot of product type j for step k; Y is the total number of processing steps for product type j; c_i is the unit cost of consumable i; $Q_{i,j,k}$ is the number of lots of product type j using consumable i for step k; $Q_{j,k}$ is the total number of lots of type j that pass inspection at step k.

The consumable usage cost requires that the ratio of the number of lots processed at each processing step to the number of lots that pass inspection is available. This ratio is a statistic that must be generated by the simulation and accessed by the cost analysis system.

(c) Product-allocated cost

This is the cost of each good product lot of all types. It is computed as the sum of the machine usage, operator usage and consumable usage costs. It can be formulated as

$$S_j = M_j + O_j + C_j \tag{1.5}$$

where S_j is the total base production costs of all good lots of type j; M_j is the machine usage cost for product j; O_j is the labor usage cost for product j; C_j is the consumable usage cost for product j.

The formulations listed above are used to estimate the base production cost of the products. Raw material costs do not vary through the production process, and are therefore not included in the formulations. Since the cost of scrapping a lot at each processing step is allocated to the good lots that pass through that step, the product allocated cost does not include a separate formulation for the scrapped lots. If the defective lots were reworked, then some of the scrapped lots may pass inspection the second time. In such a case, an appropriate adjustment must be made to the scrap costs to reflect the cost of inspection after rework. The framework of the cost analysis system includes these cost-modeling functions to estimate the base manufacturing cost of the products being manufactured.

1.3.3 Issues related to the chosen simulation language

Since the cost analysis system is designed outside the simulation language, it is necessary for the statistics generated by the simulation to be written to databases that are accessible by the system. As mentioned before, statistics about the time a product used a resource (machine or operator) must be captured and this time must be apportioned over the good lots that pass inspection at each processing step. This value, along with an identifier, has then be to written to a database, for each of the products. In this case each record would contain an identifier and the average of the tally variable. The identifier can be arbitrarily chosen, but care should be taken to see that they are consistent in both the simulation model and the cost analysis system. This is because the tally values are referenced by the system with the help of the identifier. This can be illustrated in the form of the following pseudo code.

1. Schedule the arrival of a product lot at the activity center.
2. Place the lot in a queue if the resource is busy.
3. Capture the resource.
4. Increment counter for each lot that is processed.
5. Delay for processing by the resource.
6. Release the resource.
7. Tally the processing time.
8. Dispose the entity if it fails inspection.
9. Increment counter for each lot that passes inspection.
10. Average processing times over the number of entities that pass inspection.
11. Write the product number and average processing times into a file.
12. Write the ratio of the number of lots processed to the number of lots that pass inspection at each processing step into a file.

The steps described above can be modeled in many of the commercially available simulation languages. SIMAN has been chosen as the simulation language in this effort due to its wide usage and the capability to report a variety of statistics.

1.3.4 Conceptualization of a framework for the cost analysis system

Based on the issues and requirements discussed in the previous sections, a framework for the simulation-based cost analysis system was conceptualized. The core of the framework is a control module which invokes other modules, as and when required. The control module, through an interface, prompts and leads the user through the entire cost analysis process. The activity-based costing functions are available to this control module, through the activity-based costing module. The cost and production parameters are written to static databases, and the simulation statistics are written to dynamic databases. Both set of parameters are updated through the control module. The control module runs various user-interface modules to lead the user through the process of updating the information. Once all the pertinent information is in the databases, and the simulation model is updated by the user, the cost analysis system can be invoked by the user. When the system is invoked, the simulation module is executed, which in turn runs the simulation and results in the updating of the databases containing the activity times. The activity-based costing modules are then invoked by the control module which uses the cost modeling functions and the information stored in the databases to estimate the base manufacturing cost of the products. This estimate is then conveyed to the user through the user interface. Figure 1.1 shows the framework for a simulation-based cost analysis system.

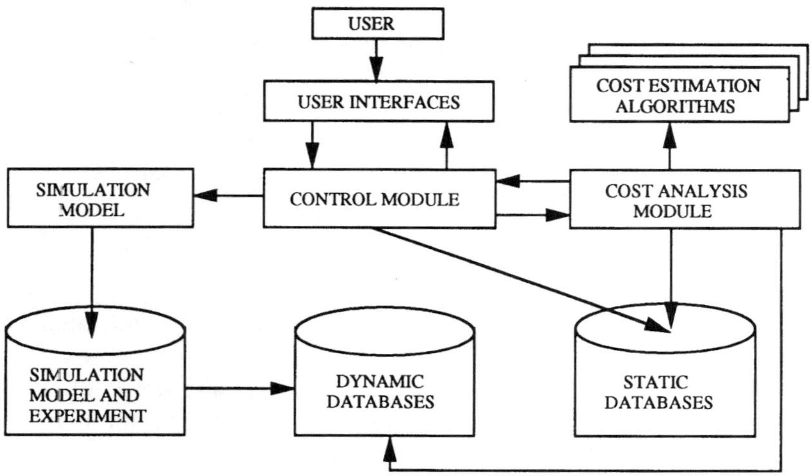

Fig. 1.1 Framework for a simulation-based cost analysis system.

1.4 SYSTEM DEVELOPMENT AND IMPLEMENTATION

For the purpose of verifying the feasibility of the simulation-based cost analysis system, a prototype system has been developed and implemented. The framework discussed in the previous section has been implemented using Turbo C++ and SIMAN and implemented on an IBM 386PC with DOS4.0 operating system. Both the hardware and software were selected on the basis of their ease of availability and portability. SIMAN was used to model a hypothetical manufacturing scenario for testing the prototype cost analysis system. The functions of the various modules are discussed in the following sections.

1.4.1 The control module

The control module allows the user to select the database to be initialized or modified and calls the relevant user interface to lead the user through this process. The control module also has access to the ABC functions which are used to estimate the base manufacturing cost. Once the user completes updating the database, the control module runs the simulation module, and uses the statistics generated by the simulation to estimate the base manufacturing cost. The operational sequence of the control module is shown in Fig. 1.2.

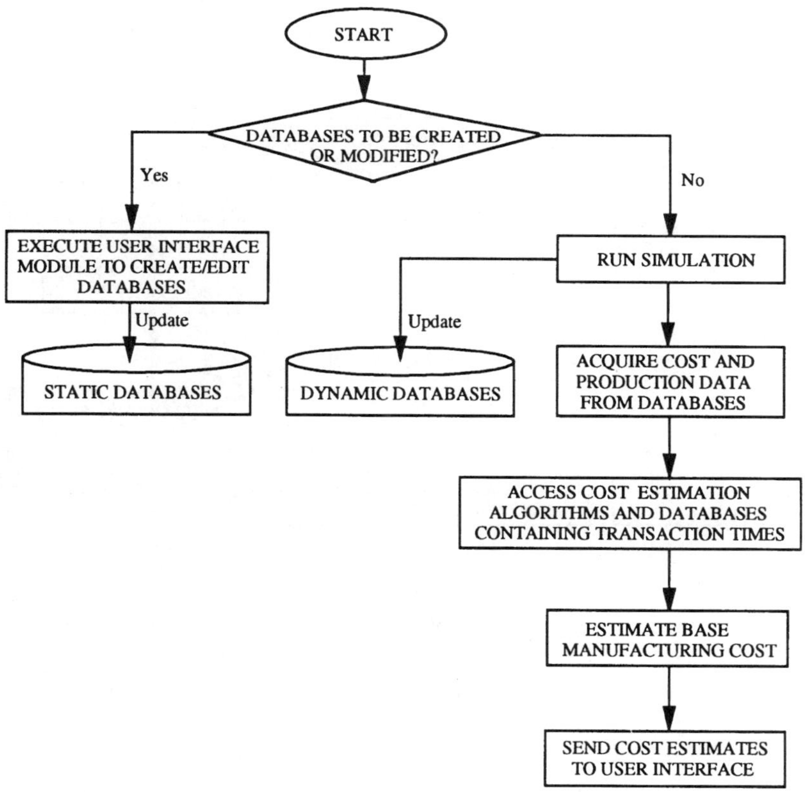

Fig. 1.2 Operational sequence of the control module.

1.4.2 The user interfaces

The user interfaces are basically menu screen and displays, and they prompt the user for the parameters necessary to update the databases containing cost and process flow information. Each of the resource categories (machines, labor and consumables) has two sets of user interfaces. The first one prompts the user about the cost information for that particular resource, and the second prompts the user about the process flow information. These parameters are then written to the relevant databases which are then accessed by the control module to

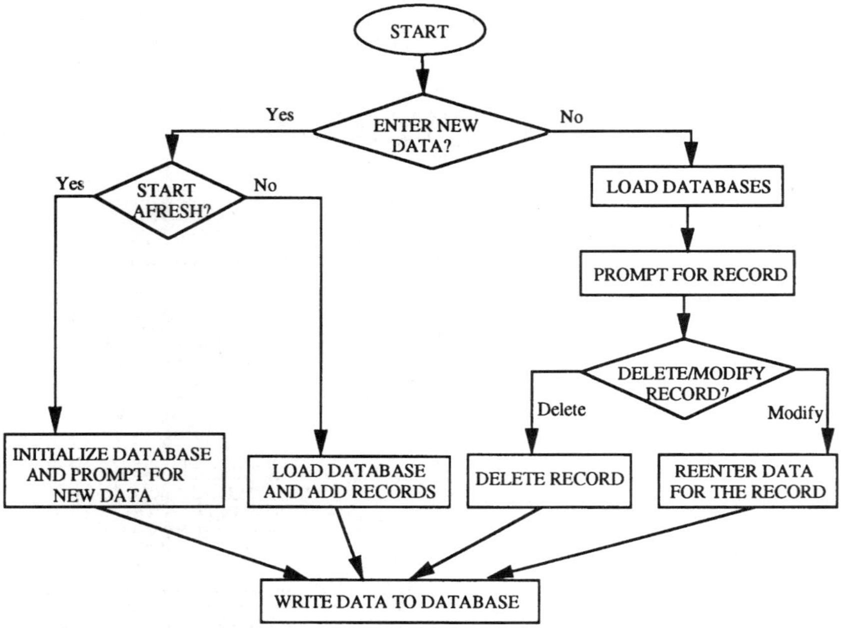

Fig. 1.3 Operational sequence of the user interfaces.

estimate the production cost. The user interfaces also allow the user to add new information or to delete existing information. Figure 1.3 shows the operational sequence of the user interfaces module in the cost analysis system.

1.4.3 The simulation module and the simulation model

The simulation module is invoked by the control module after all the cost and production parameters have been entered by the user. The simulation module prompts and receives the SIMAN file name that is to be run. The simulation module also allows the user the choice of using existing simulation statistics to estimate the base manufacturing costs instead of running the simulation to collect statistics. This would be the case when only the cost parameters have been changed and the simulation statistics are still valid. Figure 1.4 shows part of a sample generic station submodel developed in SIMAN which illustrates how the required cost parameters can be recorded in the model and output to files. The logic of the model is simple and generic, and is as follows. Product lots are scheduled to arrive at stations and enter a queue to seize a resource. If the resource is available it is seized and the simulation time is incremented by an amount equal to

```
STA1    STATION,   MACHINE1;               !Station ID
        QUEUE,     MACHINE1Q;              !Queue for Machine 1
        SEIZE:     MACHINE1R;              !Seize Machine 1
        COUNT:     LOTS;                   !Count Entering
                                           !Lots
        DELAY:     0:MARK(CAPTURE);        !Mark Time Now
        DELAY:     PTIME;                  !Delay for
                                           !Processing Time
        RELEASE:   MACHINE1R;              !Release Machine 1
        TALLY:     M1P1, INT(CAPTURE);     !Tally Processing
                                           !Time

        BRANCH,1:
                   WITH,0.1,SCRAP:         !Reject 10% of the
                                           !Lots
                   WITH,0.9,GOOD;          !Accept Remaining
                                           !Lots
SCRAP   DELAY:     0:DISPOSE;              !Dispose Off Scrap
GOOD    COUNT:     GOODPARTS;              !Count Good Parts
        WRITE,     M1:                     !Write Output to
                                           !File
          NS,TAVG(M1P1)*TNUM(M1P1)/
          NC(GOODPARTS):DISPOSE;
```

Fig. 1.4 Generic station submodel developed in SIMAN.

the processing time. The resource is then released. The cost analysis system requires that the time required for processing at each of the resources is captured and reported. This is done using the functions available in SIMAN and requires only a minimum amount of additional coding.

In the model, the attribute CAPTURE is set to the value of the simulation time at the point the resource is captured, the entity seizes the resource. The simulation time is then delayed by an amount equal to the processing time of the part, and the resource MACHINE1R is then released. The tally variable MIP1 is then used to tally the time that the lot has captured the resource MACHINE1. At this point 10% of the lots are scrapped, and the lots that pass inspection are counted using the counter GOODPARTS. The functions TAVG (tally variable) and TNUM (tally variable) are used to calculate the product of the average of the tally variable and the number of entities that updated the variable. This value would be the sum of the transaction times of a particular product type that used the resource MACHINE1. This time is then apportioned over the good lots that pass inspection. The number of good parts is accessed by the function NC (counter-name). This value is then written to the file (whose identifier is M1) along with an attribute which is the product type identifier (in this case the attribute NS). Each record of the file would then contain an identifier and the average of the tally. Care should be taken to use distinct tally variables for each product type at each resource. The cost analysis system accesses this database by matching the product numbers, and using the tally average associated with it to drive the time-based costs. As the material flow information is entered for each of the products (through the user interface), the products are given default product numbers in the order in which they are entered. The SEQUENCES element in SIMAN is used to describe the process flow of the products in the system.

1.4.4 The static and dynamic databases

The static databases maintain parameters such as machine information (purchase cost, installation cost etc.), operator information (operator identification (ID), wages/hour etc.), consumable information (consumable ID, cost/unit etc.), which are not modified or generated by the simulation. The dynamic databases maintain simulation-generated statistics, such as processing time and consumable usage. The static databases are accessed through the user interfaces for the user to enter, edit or delete information. The dynamic databases are accessed by the control module and used by the cost estimation algorithms.

1.4.5 The cost analysis module

The formulations of the various costs which were discussed earlier are implemented in the cost analysis module. The transaction times of the product lots at the various resources and their consumable usage are tallied by the simulation model and written out to the dynamic databases. The control module accesses these processing times and makes them available to the cost analysis module which then estimates the base manufacturing cost of the products. This cost is then sent to the user interface and displayed to the user. In this work, the design of the cost analysis module is restricted to the use of only those cost modeling functions discussed earlier. However, the cost analysis module can be enhanced further to include additional functions and other cost analysis techniques.

1.5 SYSTEM VALIDATION

For the purpose of validating the cost analysis system, a hypothetical manufacturing scenario was formulated. The base production cost was estimated for this manufacturing scenario using both the simulation-based cost analysis system (i.e. off-line cost estimation) and by the simulation model with the cost estimation statements built-in (i.e. on-line cost estimation) the model. The cost analysis system was validated by comparing the estimates of the base manufacturing costs computed by the off-line system with the results of the on-line system.

1.5.1 Validation scenario

In the hypothetical manufacturing scenario, three types of products are produced, J1, J2 and J3. The jobs arrive with an equal probability, and follow an arrival distribution which is exponentially distributed with a mean of 5. The process flows for the three products and their processing times are shown in Table 1.1. Four types of machines, M1, M2, M3 and M4

Table 1.1 Process flow for the three products used in the validation scenario

Product	*Process step*	*Machine ID*	*Operator ID*	*Processing time for step*	*Consumable ID*	*Units of consumable*
J1	1	M1	O1	5	C1	2
	2	M2	O2	6	C1	1
	3	M3	O3	7	C2	1.5
	4	M4	O1	8	C2	0.5
J2	1	M1	O1	5	C1	1
	2	M2	O2	6	C2	0.5
	3	M4	O3	7	C2	2
	4	M3	O2	8	C1	1.5
J3	1	M1	O1	5	C1	1.5
	2	M2	O2	6	C2	1
	3	M4	O3	7	C1	2
	4	M1	O3	8	C2	0.5

Table 1.2 Cost parameters of resources used in the validation scenario

Machine ID	*Purchase cost ($)*	*Installation cost ($)*	*Maintenance cost*	*Life (years)*	*Working hours/year*	*Consumable ID*	*Cost/unit*	*Operator ID*	*Wages/ hour ($)*
M1	100000	1000	3000	10	8000	C1	$20	O1	20
M2	150000	2000	3000	10	8000	C2	$30	O2	30
M3	120000	1500	2500	10	8000			O3	25
M4	200000	3000	5500	10	8000				

are used in the production process. One percent of all lots are rejected after each processing step. Three types of operators and two types of consumable are also used. The usage of labor resources and consumables is also shown in the table. The cost information for the machines, labor resources and consumables is shown in Table 1.2.

1.5.2 Analysis of results

The simulation model was run for both the off-line and on-line models of the hypothetical manufacturing scenario and the results were compared. The results were found to match exactly as shown in Table 1.3, thus validating the feasibility of the implemented cost analysis system. As demonstrated through the implemented prototype cost analysis system, the off-line estimation of costs has several advantages over accumulating the costs within the simulation itself. The developed system considerably reduces the additional modeling that will be necessary if the cost parameters were built-in within the simulation model. The system also has the flexibility to allow users to change the process flow easily by leading the user through the necessary steps for modifying the databases. Users can also play 'what if' scenarios by changing the cost parameters and can evaluate the impact of the changes on product costs. The off-line analysis of costs reduces the number of attributes required in the model, and as a result, the memory required to run the simulation. The

Table 1.3 Comparison of costs estimated by on- and off-line techniques

Product ID	*Cost estimated off-line ($) (cost analysis system)*	*Cost estimated on-line ($) (Cost built-in the model)*
J1	744.34	744.34
J2	826.33	826.33
J3	954.64	954.64

developed system being a prototype does not have many of the features that a commercial system would require. The need for further development is apparent and some of the features that can be incorporated in the future to enhance the system are recommended in the next section.

1.6 CONCLUSIONS AND RECOMMENDATIONS

1.6.1 Conclusions

Many managerial decisions are based on whether the benefits of a proposed change outweigh the costs associated with it. Traditional simulation models help in the estimation of performance measures, such as machine usage and manufacturing lead times. However, to get a better insight into the impact of proposed changes in a manufacturing system, it is necessary to know the implications of the changes on product costs. Simulation results can be better used to make decisions if they can also provide information about the financial bottom line. A review of past work in this area indicates that there is a lack of unified approach to interfacing costing with manufacturing simulation. In this research, a prototype simulation-based cost analysis system which uses activity-based costing techniques to estimate the base manufacturing cost of the products has been designed developed and implemented. The developed system uses the results generated by the simulation to analyze and arrive at an estimate for the base manufacturing costs of the products manufactured. The cost analysis system is maintained user-transparent and the user does not have to make major modifications to the simulation program to use the system. The system leads the user through the various steps for transcribing the manufacturing scenario and entering the required input data into the system. Precious computing power is also saved because of the elimination of additional attributes that would have been needed to model the cost parameters within the simulation model itself. The other advantage of using an off-line system is that existing simulation statistics can be repeatedly used to experiment with the effect of changes in the cost of activities performed.

1.6.2 Recommendations

The feasibility of developing a simulation-based cost analysis system to aid in manufacturing decision making was explored in this research. However, the developed system, being a prototype, has certain limitations that can be resolved in the future. One of the major limitations of the implemented prototype system is the simplification of the product cost structure. The implemented system does not take into account the indirect costs, and this is a serious limitation. The current system is capable of including the indirect costs, if the activities that are responsible for these costs can be captured in the simulation model. However, from a simulation standpoint this may be cumbersome and unnecessary, and further, the allocation of some indirect costs are also controversial in nature. For example, several different approaches are currently being followed in the industry to allocate the cost of machine and labor idle times. In a generalized decision support system, it would be inappropriate to force the user to follow one particular strategy in allocating these idle time costs. One solution may be to identify and design a number of indirect cost allocation strategies and give users the choice of selecting a preferred strategy from a list of possible strategies.

The developed system estimates base production costs for each run of the simulation separately. To develop a confidence interval of the product-allocated costs, the simulation would have to be run for multiple replications. This is a limitation that warrants further exploration. Time spent by the products as 'work in process' also adds to the manufacturing cost, but this cost has not been included in the estimation of base manufacturing costs, even though work-in-process times are available from the statistics generated by simulation. This feature can easily be added in a full scale implementation. The estimation of scrap costs has been greatly simplified in this research, and this should be enhanced to include the cost of both rework and rejects. The activity-based costing algorithms developed in this effort can also be expanded to provide a greater level of detail. Ideally, the cost analysis system should also include an expert system module to help the identification of value-added and non-value-added activities and suggest suitable modifications to the process. The prototype system currently interfaces only with the simulation language SIMAN, but in a full-scale implementation this restriction can be eliminated by allowing the user to choose a simulation language from a list of allowable simulation languages.

ACKNOWLEDGEMENTS

The authors would like to thank the Society of Manufacturing Engineers at Dearborn, Michigan, for its partial funding of this effort through the Research Initiation Grant No. 592–1723, awarded to the Department of Industrial Engineering at Northern Illinois University for the period 6/1/1992 to 6/30/1993.

REFERENCES

Beaujon, G.J. and Singhal, V.R. (1990) Understanding the activity costs in an

activity-based cost system. *Journal of Cost Management*, Spring, 51–72.
Christy, D.P. and Kleindorfer, G.B. (1990) Simultaneous cost and production analysis of manufacturing systems, in *Proceedings of the 1990 Winter Simulation Conference*, Society for Computer Simulation, New Orleans, Louisiana, USA, December, pp. 582–89.
Cooper, R. and Kaplan, R.S. (1987) The two-stage procedure in cost accounting: part one, in *The Design of Cost Management Systems*, Prentice-Hall, Englewood Cliffs, New Jersey, pp. 147–56.
Haeme, R.A., Huttinger, J.L. and Shore, R.W. (1988) Airline performance modeling to support schedule development: an application case study, in *Proceedings of the 1988 Winter Simulation Conference*, Society for Computer Simulation, San Diego, California, USA, December, pp. 800–05.
Kaplan, R.S. (1989) Management accounting for advanced technological environments. *Science*, August 25, 819–23.
Ketcham, M.G., Shannon, R.E. and Hogg, G.L. (1989) Information structures for simulation modeling of manufacturing systems. *Simulation*, **52**(2), 59–67.
McLanahan, C. and Ketcham, M.G. (1990) The representation of accounting data in industrial simulation, in *Proceedings of the 1990 Winter Simulation Conference*, Society for Computer Simulation, New Orleans, Louisiana, USA December, pp. 570–77.
Moore, R.D. (1990) Effective cost modeling on the factory floor: taking simulation to the bottom line, in *Proceedings of the 1990 Winter Simulation Conference*, Society for Computer Simulation, New Orleans, Louisiana, USA December, pp. 590–94.
Muller, D.J., Jackman, J.K. and Fitzwater, C. (1990) A simulation-based work order release mechanism for a flexible manufacturing system, in *Proceedings of the 1990 Winter Simulation Conference*, Society for Computer Simulation, New Orleans, Louisiana, USA December, pp. 599–602.
O'Loughlin, M.J. (1990) Financial simulation: combining cost information in systems analysis, in *Proceedings of the 1990 Winter Simulation Conference*, Society for Computer Simulation, New Orleans, Louisiana, USA December, pp. 578–81.
Pegden, D.C., Shannon, R.E. and Sadowski, R.P. (1990) *Introduction to Simulation Using Siman, McGraw-Hill*, Highstown, New Jersey.
Whitehouse, G.E. (1973) *Systems Analysis and Design Using Network Techniques*, Prentice-Hall, Englewood Cliffs, New Jersey, pp. 150–70.

CHAPTER TWO

A decision support system for the justification of computer-integrated manufacturing

Suleiman K. Kassicieh, H.V. Ravinder and Steven A. Yourstone

2.1 INTRODUCTION

American companies are currently seeking ways to regain their competitive edge, market share and reputation for quality that they have lost over the last two decades. Their efforts have been aimed at:

1. improving times of response to rapidly changing customer needs and tastes by improving flexibility of operations;
2. bringing quality up to Japanese levels;
3. achieving these at lower costs.

Several writers from both industry and academia have argued that to achieve these goals in the current competitive climate a revolution is needed in manufacturing thinking, a revolution that will produce the '**factory of the future**' (Meredith, 1987; Jelinek and Goldhar, 1984). Such a factory will be fully computer integrated. **Computer-integrated manufacturing** (CIM) is defined as the logical organization of individual functions – engineering, production, marketing and other support functions – into a computer networked system which combines newer **manufacturing technologies** such as flexible manufacturing systems (FMSs) and computer-aided engineering, design and manufacturing (CAE/CAD/CAM) with Japanese ideas of production control like 'kanban' and 'just-in-time'.

Any firm considering the technological leap from conventional manufacturing to CIM must justify, to itself and to its stakeholders, the considerable amount of money, time and effort needed (e.g. Daugherty, Parsaei and Kolli, 1993; Parsaei, Wilhelm and Kolli, 1993; Kassicieh, Ravinder and Yourstone, 1993). The analysis of an investment in a CIM facility poses difficulties that stem from the fact that the investment is not simply in a new piece of equipment but in a totally new way of manufacturing that will affect, fundamentally, every aspect of the operation of the firm. Typically, firms which have successfully implemented CIM report:

1. a thorough re-evaluation of the firm's goals and organization,
2. reduction of design–engineering–production–marketing lead times;
3. quality improvements.

These have led to a repositioning of the firm in the market it serves and even a redefinition of this market.

Traditional investment analysis methods have difficulty incorporating such changes (French, 1984; Weston, 1988; Canada, 1986; Attaran, 1989; Noble, 1989; Meredith and Suresh, 1986) because typically, these techniques require that all inputs be quantified. The timing and amount of cash inflows and outflows over the time horizon are estimated by some means, and the time horizon itself is selected to be some fixed number of years. The discount rate is set by management judgment to reflect the cost of funds to the firm, plus a desired rate of return on the investment, plus some adjustment for the level of risk involved. Kulatilaka (1984) discusses the determination of an appropriate discount rate for incorporating the systematic risk associated with each cash flow.

The particular financial tool (e.g. payback period, internal rate of return, net present value) used to evaluate a set of alternative investment proposals will take some or all of these factors into account (Higgins, 1977; Schall and Haley, 1980; Canada and Sullivan, 1989). The difficulty with most of these financial tools is that the decision is based on a single figure of merit (e.g. payback period, internal rate of return, net present value), and this single figure is developed from incomplete and imperfect information. Clearly, the justification problem is far too complex for a single number to be adequate for the purpose. Therefore, an elaborate sensitivity analysis may be needed to understand fully the behavior of this single measure.

Furthermore, traditional investment analysis is biased in favor of short-lived projects, the costs and benefits of which are easily quantified, and against CIM-like projects, where benefits accrue over the long term and are harder to quantify. Extra care must therefore be taken when applying discounted cash-flow techniques to CIM projects. For the purpose of capturing these costs and benefits, this chapter suggests that traditional investment analysis tools should be supplemented by other approaches like optimization and simulation.

This chapter is organized as follows. First, it describes systems used for CIM justification providing information about the elements of a proposed **decision support system** (DSS). It then discusses the weaknesses of traditional accounting methods in measuring the costs and benefits associated with CIM and suggests a solution in the concept of activity-based costing. The chapter briefly discusses the costs and benefits associated with CIM that are most likely to affect a company's profit and loss statement. This is followed by a detailed discussion of the optimization and simulation modules in the DSS. Finally, a step-by-step description of how the DSS might operate and a discussion of further research in this area is presented.

2.2 A DSS FOR CIM JUSTIFICATION

Monahan and Smunt (1987) have developed a multilevel DSS for justifying conversion to automated flexible manufacturing. Their DSS is based on three levels of analysis:

1. the long-term, firm planning level;
2. the medium-term, factory operating level;
3. the automated-work-cell-design level. At level 1 they formulate the acquisition problem as a finite-state Markov process against the criterion of cost minimization. In a follow-up paper (Mohanan and Smunt, 1989) they incorporate interest rate changes as well as the uncertain costs of **flexible automation** in their finite state Markov chain model. At levels 2 and 3, two commercial software packages, PLACE (a component of McDonnell Douglas Information Systems Group Software for use in work-cell design) and XCELL for work-cell simulation were used. They begin with simulation and end with a Markov chain optimization model. However, the issue of difficult-to-quantify costs and benefits is not addressed in their paper.

Our hierarchical approach, combining optimization and simulation tools, avoids the inherent weaknesses of these two separate approaches to modeling a CIM system. Mixed-integer programming cannot incorporate uncertainties or detailed precedence relationships among jobs without becoming computationally intractable. Simulation will not directly generate alternative job–machine configurations, but can evaluate proposed configurations – generated by the mixed-integer program. The evaluation begins with a mixed-integer optimization model to identify the optimal mix of traditional and flexible machines. Simulation is then employed to study the performance of the optimal mix. Several iterations through the optimization and simulation phases may be required in order to study the effects of different assumptions and inputs. Finally, we include a rating scheme to incorporate difficult-to-quantify costs and benefits of CIM.

The contribution of this chapter is to combine, in a model-based DSS the appropriate accounting, optimization, simulation and decision-making techniques that are required to evaluate the change from traditional (i.e.

non-CIM) environments to varying levels of CIM implementations. In order to generate realistic cost information for the purpose of conversion to some level of flexible automation, a new approach to cost accounting is required. The importance of activity-based cost accounting will also be discussed in section 2.4. The user of the proposed DSS may also generate different scenarios based on different levels of demand as well as costs, and then perform extensive sensitivity analyses.

2.3 DSS DESCRIPTION

The overall structure of the DSS is shown in Fig. 2.1. The DSS allows the decision maker to use information from a variety of sources to determine the major costs and benefits that need to be examined. Where the use of models is not feasible due to complexity of the situation being modeled or absence of data, the user can subjectively estimate the inputs or parameters needed. The **dialog** subsystem links the decision maker to both the **database** and the **modeling** subsystems. The database subsystem is loaded with the detailed information about the organization's external and internal performance measures. The modeling subsystem provides the decision maker with mathematical, financial, simulation and accounting models.

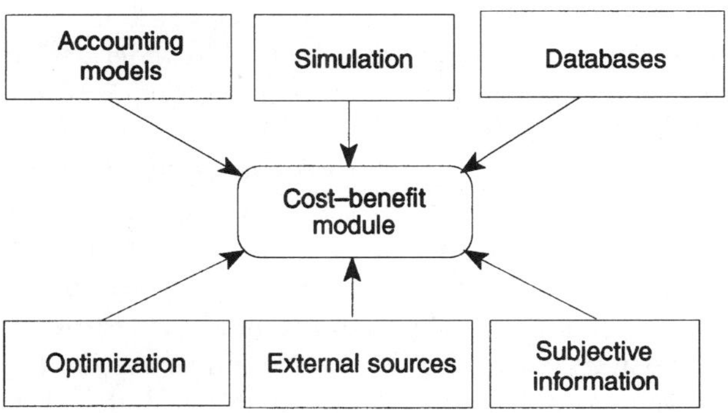

Fig. 2.1 Overall structure of the DSS depicting inputs to the cost–benefit module of the justification process.

Specifically, the DSS incorporates:

1. **activity-based costing** (*ABC*) to reflect a better allocation of costs to activities;
2. **mixed-integer programming** (MIP) to determine the number of machines to invest in, labor costs, and floor space required;

3. **simulation** and other **optimization models** to incorporate uncertainty and precedence relationships, and to provide an assessment of the proposed configuration.

In the following sections we will discuss these and related topics in detail.

2.4 ACTIVITY-BASED COSTING

Kaplan (1983, 1986, 1988, 1990) points out that accounting systems have not been modified to reflect the true nature of manufacturing activities performed in new technological environments such as CIM. In innovative and successful firms, accounting systems are more than simply reporting systems. Traditional accounting allocates fixed costs and overheads to a particular product as a percentage of the total direct labor hours allocated to that product. With highly automated manufacturing systems, direct labor is no longer the dominant input resource. Other measures of activity have to be used in addition to labor hours. This is the motivation for *ABC*. Hiromoto (1988) concludes that the accounting techniques used by many Japanese firms help to clarify their organizational goals, in addition to their traditional function of reporting externally on their operations. For a firm to gain knowledge of the true costs of delivering products to its customers, accounting systems must reflect the realities of automated manufacturing systems such as the substitution of machinery for labor, the closer ties between design, manufacturing and marketing geared towards delivering products demanded by the customer and the ability to anticipate and respond rapidly to changes in consumer needs.

Cooper (1990a,b) suggests using the number of set-ups incurred in the manufacture of a particular product, the distance that material is moved by material-handling equipment in the course of manufacture of that product and the number of components and subassemblies used in the product. There are other factors that could be used in determining the best variables to use to allocate overhead to product cost. These factors could include issues such as batch size. Batch size influences the time between switches from one set-up to another. The smaller the batch size, the higher the ratio of set-up time to total manufacturing time per unit produced and the higher the allocation of overhead to products produced in smaller batches.

ABC satisfactorily addresses some of the problems associated with using traditional accounting techniques to justify CIM. However, in comparing current manufacturing capabilities with proposed CIM acquisitions, it might be necessary to transform the current accounting system to ABC so that a fair comparison of costs under the old system and the new one would be possible.

No accounting method, however modified, can address strategic issues like increased competitiveness and changed market perception of the firm, or the creation of a new market niche, and their effect on the survival of the firm. These crucial long-term issues have to be addressed by other, albeit subjective, methods.

2.5 CIM AND THE FIRM'S PROFIT AND LOSS STATEMENT

Any justification effort will require an exhaustive evaluation of the impact of the proposed CIM system on the firm's operations now and in the future. A typical profit and loss statement will have the components shown below:

- sales revenue
- labor cost
- material cost
- overhead cost
  - supplies
  - utilities
  - repairs
  - maintenance
  - indirect labor
  - depreciation
  - rent
  - insurance
  - management.

As can be seen from the above list, overhead costs are an important issue in costing products. If ABC is to be implemented, the information system in the plant must track some measurements that are not traditionally collected. For instance in traditional accounting systems, supplies are allocated to products as a fixed percentage related to the labor component in the product. The modified information system should be designed to track supplies in each product and use direct costing to control supplies and to allocate cost correctly. Other overhead costs require the collection of new information. Rent might be allocated to products based on the percentage of time each unit is machined. This requires tracking of machining time for each product but with automated manufacturing, time in and out of a machine is not very hard to track. These calculations require a change in the way that information is collected about the activities in the plant but they are necessary for the determination of the true cost of each product. If these numbers have not been captured for past data, the decision maker might want to estimate these parameters so that the allocation of overhead costs follows the same rules for past and future performance.

Once the **costs** and **benefits** of a particular CIM configuration are quantified, it will be possible to construct a projected profit and loss statement for the firm's operations with that configuration. If several CIM configurations are being considered, several such profit and loss projections may have to be made. These projections are then compared with profit and loss projections assuming that the firm continues to operate with its existing traditional manufacturing systems. The information in the profit and loss statements is arrived at as follows.

2.5.1 Sales revenue

Forecasting future sales of established products could follow traditional time series methodology, whereas market research might be needed for new products. The decision maker needs to get these data from the studies conducted by the marketing team of the organization. Firms that implement CIM successfully gain considerable long-term strategic advantage which can dramatically increase sales and contribution margins. While difficult to quantify, these advantages result from:

- reduction in throughout times;
- increased capacity;
- reduction in engineering change order implemention time;
- increased productivity due to faster design and better process planning;
- repeatability of production plans over long and short production runs;
- improved precision and quality;
- enhanced opportunity for further automation in the future;
- increased competitive advantage;
- flexibility and sensitivity to market;
- less machine obsolescence due to flexibility in equipment;
- increased employee morale due to elimination of dirty, dull and dangerous jobs;
- improved company image;
- valuable experience to employees;
- increased innovation which outpaces competitors' manufacturing capabilities.

2.5.2 Costs

The number of new machines and their cost is a central issue in revenue and cost projections. Through the use of optimization and simulation we can determine the optimal number of machines necessary for producing a specified bundle of products. For example, once forecasts of the type and volume of products that are needed have been made, the optimal mix of flexible and traditional equipment, the number of labor hours required to operate this mix, as well as the cost of installation and operation of machinery are determined in our DSS by a mixed-integer program. Through simulation we then evaluate the feasibility of the proposed configuration and layout of machines; several iterations may be necessary to obtain a satisfactory arrangement. As a by-product of the simulation we obtain estimates of production and handling costs, throughput times, quality levels, work-in-process (WIP) and floor-space requirements.

(a) Direct labor and material costs

Direct labor costs can be derived from the number of machines, their capacity and need for labor to produce the products to meet projected demand. Material cost may be estimated as a function of demand. Simulation can also yield estimates of both these costs thus allowing checking for consistency.

(b) Overhead costs

These costs include the following.

1. **Floor-space**: the implementation of CIM might result in savings in the annual investment in and/or rent of the required floor-space. The number of machines, work cells, storage and other work areas required in the new environment will determine the floor-space required and consequently the rent paid. The saving in floor-space, if any, is determined from the mixed-integer program and the simulation that follows. Within limits, rent may be assumed to be directly proportional to space used. However, it is not always clear that the organization can use any marginal saving in space for other productive purposes unless the saving is large enough.
2. **Utilities**: these and other variables dependent on space are calculated in the same fashion as above. Power requirements per machine could be calculated from specifications. Lighting per unit area, water needs for production as well as other utilities can be calculated depending on the set-up of the new plant and equipment. Allocation to different products will be done using ABC. It would be beneficial to keep the estimates obtained during the planning stage to compare to the actual values realized to help in setting up the ABC system.
3. **Indirect labor**: cost reductions here might be possible by the extensive substitution of automatically guided vehicles in the proposed CIM system for manual labor in material handling, and by the reduction of WIP and finished goods inventories.
4. **Other costs**: costs such as supplies, repairs, maintenance, depreciation on equipment, taxes, insurance and management and supervisory costs are calculated using the production volume figures, the number of machines or other related variables.

2.5.3 Benefits

The benefits of the CIM system can be divided into three major groups: the first set of benefits will be reflected in the increase in sales and revenues, the second in lower costs. The third group consists of the intangibles. In comparing sales and costs in the past to projected sales and costs in the future under the different CIM options, the decision maker needs to consider the intangibles even though they are hard to quantify. The decision maker can estimate a financial figure for each one of the intangible benefits or use a ranking scheme to compare the different options.

Table 2.1 summarizes the impact of CIM on each of these components and suggests sources of information from which they may be estimated. Where no feasible sources exist, costs and benefits may have to be estimated subjectively.

2.6 OPTIMIZATION MODELS IN CIM JUSTIFICATION

Once demand has been estimated, the DSS uses mixed-integer programming to determine the number of flexible machines in which to invest, the labor costs and the floor-space required. Several other approaches have been proposed in the literature and we discuss some below.

Programming models have a rich development in the literature. Several authors have examined the relationship between CIM implementation and some, but not all, of the benefits of CIM (e.g. Buzacott and Yao, 1986). For example, the relationship between CIM implementation and flexibility has been mathematically modeled – while leaving out the impact on quality levels. Clearly the economic justification problem is complex. Hence, mathematical models necessarily ignore many elements of the problem in order to be tractable. There would be little hope of optimally solving a truly comprehensive model.

The following are some of the programming models that have been applied in the literature to the problem of selecting the optimal FMS or CIM system.

1. **Linear programming**: Chandra and Schall (1988) use the Leontief input–output model and linear programming to determine performance measures for an FMS in order to determine whether or not the system is economically justifiable. The Leontief input–output model assumes a linear transfer relation among all the interrelated parts of the system. Clearly, the disadvantage here is the inability to capture nonlinear relationships. The advantage of our DSS is that the nonlinearities can be studied through simulation.
2. **Integer programming**: Burstein (1986) develops a deterministic, two-product, dynamic, mixed-integer programming model of flexible

Table 2.1 Costs, benefits and data sources for the CIM justification process

	Current system[a]	*CIM options*[b]
Cost of the new machines	N	X,O
Cost of conversion	N	X,E
Cost of operations	D	X,E
Other costs	D	E
Floor-space	D	S,E
Labor	D	S,O,E
Work in process	D	S,O,E
Finished goods inventories	D	S,O,E
Materials handling costs	D	S,O,E
Employee turnover	D	S,O,E
Training costs	D	S,O,E
Supervisory costs	D	S,O,E
Increased capacity	N	S,E
Reduction in throughput time	N	O,E
Reductions in defect rates	N	O,E
Reductions in Engineering Change Order costs	N	X,E
Increased productivity in design	N	X,E
Increased productivity in process planning	N	X,O,E
Repeatability over long and short production runs	N	O,E
Enhanced opportunity for further automation	N	E
Long-term viability of the firm	N	E
Product/volume flexibility	N	E,O
Other benefits	N	E

[a] D = from database; N = not applicable
[b] E = to be estimated subjectively; S = calculated from the simulation results; O = calculated from optimization results; X = from external sources,

automation with technological improvements over time. The limitation of the mixed-integer-programming approach is its restricted ability to incorporate difficult-to-quantify costs and benefits.

3. **Markov models**: Monahan and Smunt (1989) formulate the decision problem of incrementally converting from a labor-intensive batch production process to flexible automation as a finite-state Markov decision process. The output of their finite-state Markov model is the optimal level of capacity to convert to flexible automation. Their criterion is the minimization of the sum of expected discounted costs incurred over a finite horizon. Monahan and Smunt incorporate cost considerations in their modeling approach but omit consideration of the impact of flexible automation on revenues.
4. **Stochastic programming**: Fine and Freund (1990) present a two-stage stochastic program as a product-flexible capacity investment model. They analyze the cost–flexibility tradeoffs involved in investing in product-flexible manufacturing capacity – while excluding benefits that affect revenues.
5. **Dynamic programming**: Kulatilaka (1988) presents a stochastic dynamic program for analyzing the optimal use of flexible automation. The value of flexibility is analyzed within the framework of a stochastic dynamic model. This paper extends the work of Monahan and Smunt (1989) and Fine and Freund (1990) in three important ways. First, Kulatilaka's approach captures the dynamic features of the decision problem. Second, it combines the optimal operating mode of the FMS with the valuation of the technology. And, finally, it specifies the value of the flexibility of an FMS.
6. **Goal programming**: these models view the factors as goals to be attained, subject to certain constraints on resources. Weights are used on goal deviations to assign relative importance to each of the selected factors (see Ignizio, 1976, for examples). The goal-programming approach, as well as the other programming approaches we have discussed, are limited in their ability to capture difficult-to-quantify costs and benefits.

The dynamic-programming approach to financial justification of CIM captures a key element of the decision problem that is not considered by the other programming models discussed above. The dynamic-programming approach models the dynamic elements of the justification problem. As an example, Kulatilaka (1988) notes the problem of switching between modes of production. Given that switching is costly, future decisions on switching will be driven by not only the realizations of demand but also by the current production mode – as well as past switching decisions.

Dynamic-programming (DP) methodology is well-suited to the financial justification process for a CIM investment since the decision on the level of investment for CIM depends not only on the current realizations of costs and revenues but also on prior decisions on the level of investment in CIM. Clearly we should view the justification process as a sequence of investment decisions taking place over some time horizon of

N years. In each future time period a decision concerning the level of additional investment in CIM must be made. We can model this investment problem using probabilistic DP with discounting. The DP **states** represent the level of investment in CIM at the beginning of the decision period. The DP **stages** represent the time period in years (1,2, ... , *N*), the **decisions** represent feasible levels of investment to be selected at a given stage, and the **objective** is to maximize the expected net present value of the investment.

2.7 THE SIMULATION MODEL

Manufacturing facilities differ on many dimensions such as type of product manufactured, objectives and goals of top management, manufacturing processes, plant layout and available capacity (e.g. Kassicieh and Schultz, 1987). An innovation or idea that works in one facility, may not in another. The real usefulness of simulation is in the study of alternative system designs before they are actually implemented. In our DSS the results of the mixed-integer-programming (MIP) model are passed on to the simulation module which evaluates the feasibility of the MIP results, and develops an appropriate layout, incidentally providing several cost estimates for use in the projected profit and loss statements.

Simulation has become a standard tool, in both industry and academia, for the analysis of complex manufacturing systems. The reason for this popularity is not hard to find. Mathematical programming models seeking to describe realistic manufacturing systems are often intractable. Suri and Treville (1991) suggest the use of queuing theory for modeling manufacturing systems. Their technique is used to answer questions of capacity, bottlenecks and acquisition of capacity. With the proliferation of user-friendly packages for factory simulation like SIMFACTORY II.5, SIMAN and XCELL, to name a few, the use of simulation to investigate problems and explore proposed solutions is certain to increase.

While FMSs have been simulated extensively, few simulation models have been designed with the intention of providing input to the justification decision. In the DSS, the simulation module uses SIMFACTORY II.5 (CACI, 1991), an easy-to-use programming-free simulation package. In this simulation package factory, elements are represented by various icons which are modified by pointing and clicking. Batch and continuous processes are also easily modeled in most commercially available packages. The decision maker has a choice of a large number of statistical distributions and may even use empirical distributions if deemed more appropriate. With such packages, building a model is a simple task. Modifying the model to suit various scenarios or for what-if kinds of questions is just as simple. Many performance measures are automatically generated and facilitate comparison between an existing and an alternative system.

The following hypothetical situation was modeled using SIMFACTORY II.5. Imagine a factory organized along the following simple lines. It makes two products, A and B, and uses a product layout in their manufacture. Product A requires three steps: grinding, milling and finishing, and the machinery is arranged in this order. Product B requires four steps: grinding, turning, threadcutting and finishing, and again the machines are arranged in the required order. These details are shown in Fig 2.2.

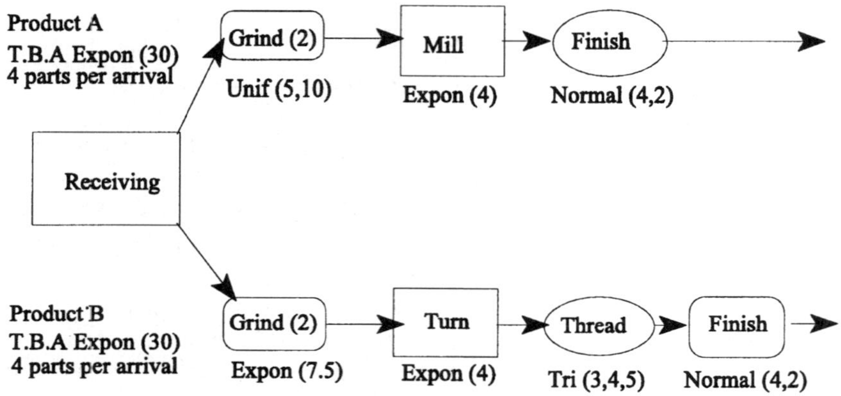

Figure 2.2 Example simulation: the existing layout. Processing times in minutes are shown alongside each operation.

Production is triggered by the arrival of raw materials. Demand for these products is such that raw material is released to the shop floor at the rate of four units every 30 minutes, exponentially distributed. Processing times are random variables, and are shown in Fig 2.2. Further, each operation is preceded by set-up and followed by tear-down, which are also random variables. For simplicity, it is assumed that set-up and tear-down are each uniformly distributed between 0.5 and 1.5 minutes at each machine. At the end of the finishing operation, some rework is discovered, which is assigned to the machines which were responsible. There is also some scrap. It is assumed that, after rework, the product is certain to be good and no more rework or scrap will be generated. A forklift truck is available to move material and WIP. Each process has a buffer before and after it.

Building a SIMFACTORY II.5 model of this situation took only a few hours. Many useful statistics were collected and maintained automatically by SIMFACTORY II.5. Considering the purpose of this simulation, the statistics of special interest are: average flow-times (or time spent in the factory by a part), machine usage, average WIP at the various buffers and the amount of material handling done by the forklift truck.

The model was run for the equivalent of seven days of one shift operation. (Statistically, this is not a large enough sample, but the purpose of this simulation is only as illustration.) The results of the simulation are summarized in Table 2.2 in the current-system column.

Table 2.2 Simulation results

(a) Output and flow times

	Current		*FMS*	
	Product A	*Product B*	*Product A*	*Product B*
Daily output	61	54	65	60
Flow time (hrs)	5.45	2.28	0.17	0.23

(b) Usage and inventories

	Current	*FMS*
Idle times	5–43%	45–55%
Set-up/tear-down	12–30%	NA
Rework finishing	10–12%	NA
Machines	2–6%	NA
Number of buffers	15	1 (receiving)
Average buffer levels	0–11 units	30
Number of forklifts	2	1
Forklift usage	5%	5%

NA: not applicable.

Now, assume that a factory is considering the acquisition of a completely automatic flexible machine which combines all five operations – grinding, milling, turning, threadcutting and finishing – in one center. The grinding operation is a high-speed operation on this machine, and its output is equivalent to four of the old ones (the proposed layout is shown in Fig 2.3).

What advantages might this machine have over the existing ones? The two products A and B might have their sequence of processing programmed into the machine, thus eliminating set-up and tear-down on the various operations. However, when switching from one product to the other, a certain changeover time might need to be allowed for.

All material handling and buffers between processing are eliminated. Further, since operations are automatic, processing times become

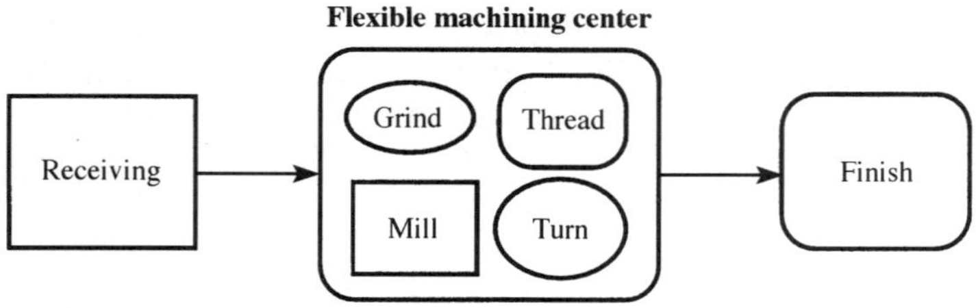

Fig. 2.3 Example simulation: an alternative layout. Processing times are the same as in Fig. 2.2 except for grinders, which are assumed to work four times faster (this assumption is made for balance).

constants rather than random variables. Finally, we have assumed that all operations produce 100% good items; there is no rework. Clearly, this assumption is not reasonable in a real CIM system. The simulation can easily be modified to study the effects of different percentages of defective items. There will continue to be some rework, scrap and downtime because of defects, since there will continue to be variability in materials, environmental conditions and machine tools. But the goal of small numbers of parts-per-million defective (e.g. 3.4 ppm) is being approached in some high-quality manufacturing systems such as Motorola's pager division. Implementation of CIM systems does not guarantee zero parts-per-million defective, but it certainly opens opportunities to achieve close to that level of quality.

Building this simulation model was even easier than the previous one. A week's operation on one shift was simulated. The results of this simulation are shown in Table 2.2 under the FMS heading. On the basis of these numbers, the proposed equipment acquisition seems to be a good idea. In the next level of the DSS, these performance measures will be converted into cash flows which will help the justification process.

An objection could be raised that the simulation example does not accurately represent a CIM factory. Monahan and Smunt (1987) demonstrate a simulation incorporating only two products run over eight machines. They state: 'Since the objective of our DSS is to make initial evaluations to justify long-term conversion strategies, precise replication of complex factories is not required. Rather, getting a basic understanding of the effect of incorporating flexible automation for numerous scenarios is most important.' It must be emphasized that the purpose of the above simple example was illustration only. A realistic simulation would be more elaborate, incorporating complex processes, several machine cells, numerous products, varying arrival patterns, order quantities and batch sizes, handling and materials control, disruptions for various reasons (breakdown, maintenance, absenteeism, shortages) and the effect of learning. Such a simulation would provide several types of measures of performance as shown in Table 2.1.

As the firm gains more experience with the acquisition of CIM technology, and as the experience of other firms becomes known, the simulation models can be enlarged to include other operations and enhanced to model the effect of CIM technology on quality levels, turnaround times for new product designs, cost of operations, product pricing and market share. These models will make the DSS a powerful tool for sophisticated users.

2.8 THE DECISION SUPPORT SYSTEM

In the above sections, the main components of the DSS were presented. The major steps of the DSS and the way they work will be briefly summarized, as shown in Fig. 2.4.

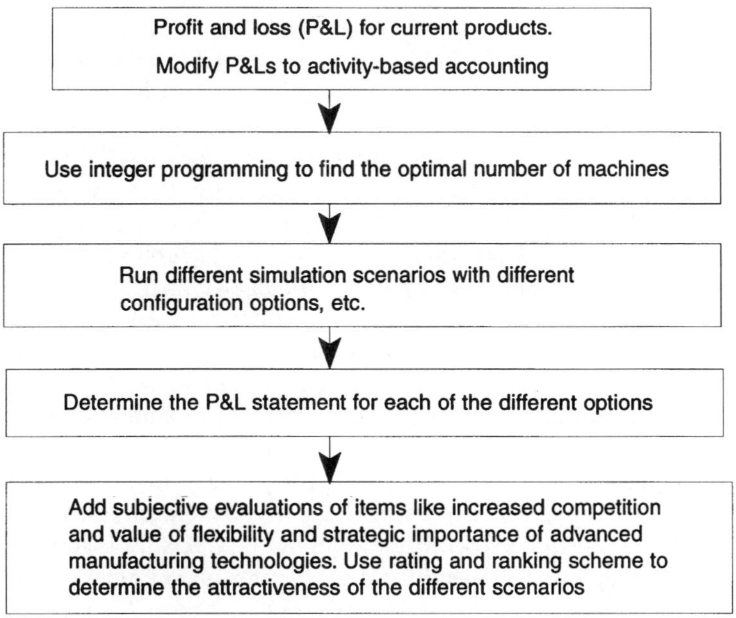

Figure 2.4 Flowchart of the DSS for CIM justification.

1. Retrieve the current profit and loss statement of the firm from the database. If one is not readily available, it may have to be constructed from the information available in the database. If, ABC is used to evaluate various proposed CIM implementations, it may be necessary

to restate the current profit and loss statement in terms of ABC. This conversion is feasible if information about cost drivers (number of set-ups, amount of material handling etc.) is available, which usually is the case. Otherwise some data collection to obtain the required information, or subjective assessment of the required data may have to be made. Once a satisfactory statement has been obtained, it is then projected into the future, assuming that traditional technologies will be used to handle planned future levels of operation, while allowing for the possibility that competitors might adopt newer technologies.
2. Use the MIP model to determine the appropriate number of machines in which to invest, the cost of such an investment as well as incremental costs of operating the machines, the labor hours required, and the cost of these labor hours.
3. Simulate the operation of the facility with the number of machines calculated in step 3 with the intention of evaluating the number and selecting an appropriate layout. It might be necessary to iterate between steps 2 and 3 several times before an operationally feasible number of machines is obtained. This might necessitate modifying the MIP model and re-solving it. At the end of this iterative process, the appropriate number of machines and a good layout are determined. In addition, several useful cost estimates are also obtained as a by-product of the simulation.
4. Construct projected profit and loss statements for each CIM option that is being considered. The figures that go into these statements are obtained from the MIP model, the simulations, existing data and subjective estimates. The time horizon over which these projections are made is an important determinant of the decisions made regarding the CIM implementations. As was noted, shorter time horizons penalize CIM projects whose real benefits become apparent only over the long run.
5. Incorporate difficult-to-quantify costs and benefits. Since quantification of all costs and benefits is not possible, a rating scheme approach to comparing different scenarios for CIM implementation has great value. Putrus (1990) has proposed using the analytic hierarchy process (AHP). The AHP is a weighted scoring model that corrects for inconsistency in human judgment. The weights for each criterion are determined by the rater by comparing criteria against one another. Clearly, a rating scheme has value for the decision maker, since not all costs and benefits of CIM will be quantifiable.

Our DSS incorporates a simple additive weighted factor scoring model. The weights are normalized so that they represent the proportion of the total weight assigned to each of the several factors. The total score for each CIM option is obtained by multiplying the weight for each factor by the CIM option's score on that factor and totaling the results.

The decision maker may then incorporate the results of the optimization and simulation models with the results of the ranking model. The DSS does not provide an optimal solution to the decision problem – only support in making an informed decision.

2.9 DISCUSSION AND FURTHER RESEARCH

The justification of an investment in CIM requires a multifaceted approach. Any one of the traditional justification approaches would necessarily fail to include all of the important factors that drive the optimal decision. Economic models, mathematical programming models, accounting models, and simulation by themselves, have limited abilities if used singularly to capture the essence of the complete CIM system. However, if they are used together, they can capture most of the important issues that need examination.

Clearly, there are significant advantages associated with the use of a decision support system. The important issues in a CIM justification problem are brought into focus when analyzed in a decision support framework. Decision makers have the advantage of studying proposed layouts of the new facility through simulation and developing better estimates of costs through activity-based accounting methods. Furthermore, the DSS allows a decision maker the opportunity to move back and forth between the accounting models, the simulation models, and the optimization model. The exploration of several scenarios, along with the associated financial ramifications, is clearly a significant benefit of the DSS approach to CIM justification.

This analysis can be augmented by further empirical research focusing on the experiences of firms implementing CIM systems, which should reveal the styles and patterns of decision making used, as well as their successes and failures.

REFERENCES

Attaran, M. (1989) The automated factory: justification and implementation, *Business Horizons*, May–June, 80–86.

Burstein, M.C. (1986) Finding the economical mix of rigid and flexible automation for manufacturing, in (eds K.E. Stecke and R. Suri) *Proceedings Second ORSA/TIMS Conference on Flexible Manufacturing Systems*, Elsevier, New York.

Buzacott, J.A. and Yao, D.D. (1986) Flexible manufacturing systems: a review of analytical models. *Management Science*, **34** (July), 890–905.

CACI (1991) *SIMFACTORY II.5* Release 3.1, CACI Products Company, 3344 North Torrey Pines Court, La Jolla, California 92037.

Canada, J.R. (1986) Non-traditional method for evaluating CIM opportunities assigns weights to intangibles. *Industrial Engineering*, March, 66–71.

Canada, J.R. and Sullivan, W.G. (1989) *Economic and Multiattribute Evaluation of Advanced Manufacturing Systems*, Prentice-Hall, Englewood Cliffs, New Jersey.

Chandra J. and Schall, S.O. (1988) Economic justification of flexible manufacturing systems using the Leontief input–output model. *The Engineering Economist*, **34**(1), 27–50.

Cooper, R. (1990a) ABC: a need, not an option. *Accountancy*, **106** (September), 86–88.

Cooper, R. (1990b) Five steps to ABC system design. *Accountancy*, **106** (November), 78–81.

Daughterty, L.C., Parsaei, H.R. and Kolli, S.S. (1993) Strategic justification of advanced automated technology systems using an expert system. *Computers and Industrial Engineering*, **25** (September), 107–10.

Fine, C.H. and Freund, R.M., (1990) Optimal investment in product-flexible manufacturing capacity. *Management Science*, **36** (April), 449–66.

French, R.L. (1984) Managements looking at CIMS must deal effectively with these issues and realities. *Industrial Engineering*, August, 70–77.

Higgins, R.C. (1977) *Financial Management*, Science Research Associates, Chicago.

Hiromoto, T. (1988) Another hidden edge – Japanese management accounting. *Harvard Business Review*, **66** (July–August), 22–26.

Ignizio, J.P. (1976) *Goal Programming and Extensions*, Heath Books, Lexington.

Jelinek, M. and Goldhar, J.D. (1984) The strategic implications of the factory of the future. *Sloan Management Review*, **25** (Summer), 29–37.

Kaplan, R.S. (1983) Measuring manufacturing performance: a new challenge for managerial accounting research. *The Accounting Review*, **58** (October), 686–705.

Kaplan, R.S. (1986) Must CIM be justified by faith alone. *Harvard Business Review*, **64** (March–April), 87–95.

Kaplan, R.S. (1988) One cost system isn't enough. *Harvard Business Review*, **66** (January–February), 61–66.

Kaplan, R.S. (1990) The four-stage model of cost systems design. *Management Accounting*, **71** (February), 22–26.

Kassicieh, S.K. and Schultz, C.R. (1987) Decision support flexible manufacturing systems. *OMEGA*, **15**, 495–503.

Kassicieh S.K. Ravinder H.V. and Yourstone S.A. (1993) Proposed design of a DSS for the justification of advanced manufacturing technologies. *IEEE Transactions on Engineering Management*, **40**(4), 398–403.

Kulatilaka, N. (1984) Financial, economic and strategic issues concerning the decision to invest in advanced automation, *International Journal of Production Research*, **22**, 949–68.

Kulatilaka, N. (1990) Valuing the flexibility of flexible manufacturing systems. *IEEE Transactions on Engineering Management*, **35**(4), 250–57.

Meredith, J.R. (1987) The strategic advantages of the factory of the future. *California Management Review*, **29** (Spring), 27–41.

Meredith, J.R. and Suresh, N.C. (1986) Justification techniques for advanced manufacturing technologies. *International Journal of Production Research*, **24**, 1043–57.

Monahan, G.E. and Smunt, T.L. (1987) A multilevel decision support system for the financial justification of automated flexible manufacturing systems. *Interfaces*, **17** (November–December), 29–40.

Monahan, G.E. and Smunt, T.L. (1989) Optimal acquisition of automated flexible manufacturing processes. *Operations Research*, **37** (March–April), 288–300.

Noble, J.L. (1989) Techniques for cost justifying CIM. *The Journal of Business Strategy*, January/February, 44–49.

Parsaei, H.R.; Wilhelm, M.R. and Kolli, S.S. (1993) Application of outranking methods to economic and financial justification of CIM systems. *Computers and Industrial Engineering*, **25** (September), 357–60.

Putrus, R. (1990) Accounting for intangibles in integrated manufacturing. *Information Strategy: The Executive's Journal*, Summer, 25–30.

Schall, L.D. and Haley, C.W. (1980) *Introduction to Financial Management*. McGraw Hill, New York.

Suri R. and Treville, S. (1991) Full speed ahead. *OR/MS Today*, June, 34–42.

Weston. C. (1988) Computer integrated manufacturing systems: fact or fantasy. *Business Horizons*, July–August, 64–68.

CHAPTER THREE

Linking strategies to actions: integrated performance measurement systems for competitive advantage

Archie Lockamy III and James F. Cox III

3.1 INTRODUCTION

International competition has changed the nature of the US manufacturing environment. Manufacturing firms must now strategically examine customers' needs in all served market segments for competitive survival. Moreover, firms must determine those products or service attributes which are 'order qualifiers' – necessary for either getting into or staying in the market place – from those attributes which are 'order winners' – necessary for increasing market share (Hill, 1989).

Order qualifiers and winners are defined by the customers in a specific market segment and may include any combination of the following:

- lower price
- better quality
- shorter lead times
- on-time delivery
- increased process/product innovation
- increased process/product flexibility
- more responsive field service.

To successfully compete globally, manufacturing firms must:

1. develop strategies to support actual customer needs;
2. use these strategies as the driving force for all organizational decisions and actions.

Strategic objectives are directives designed to affect the long-run viability of the firm (Lockamy and Cox, 1994). These objectives must be supported by specific order-winning criteria established for a particular market segment. For example, a firm may decide that increased market share is a critical strategic objective for long-term success. The firm must now determine the appropriate order-winning criteria necessary for achieving its market growth objective. When strategic objectives are realized through the order-winning characteristics of the firm's products, a competitive advantage is gained. Thus, US manufacturers need decision support systems which facilitate the development of competitive advantages in the marketplace.

3.2 LINKING STRATEGIES TO ACTIONS

A firm's strategies should provide a link between organizational objectives, functional area policies, tactical plans and operating plans. Integrated performance measurement systems facilitate linkages between strategies, objectives, policies, and tactical and operating plans by providing a means for communicating objectives and achievements throughout the firm. These systems also provide a means for assessing organizational resource effectiveness. Through integrated performance measurement systems, decision makers can determine whether resources are being used effectively in supporting overall strategy and local objectives. Also, decisions can be made as to the compatibility of local objectives and overall strategy.

Integrated performance measurement systems provide the link between the firm's resources and its customers. These systems provide information used to compare resource capability to customers' needs. Managers responsible for functional resources must determine their capability to produce products to the customers' expectations. Also, resource deployment decisions must be made to continually improve upon the current level of customer satisfaction.

Additionally, the financial impact of market opportunities must be assessed. Thus, a link between the firm's financial function and its customers is needed. Defining each customer's need and the financial impact of fulfilling the need are critical to developing a competitive advantage. Such a link is provided through integrated performance measurement systems.

A link between the firm's finances and its resources is provided through the financial and managerial accounting systems in most firms. However, problems arise when the accounting system fails to reflect the realities of the manufacturing environment. As firms adopt the philosophies of just-in-time (JIT), total quality management (TQM) and theory of constraints (TOC), in most cases the traditional accounting systems no longer provide a valid decision-making framework (Johnson and Kaplan, 1987). Integrated performance measurement systems provide the means for

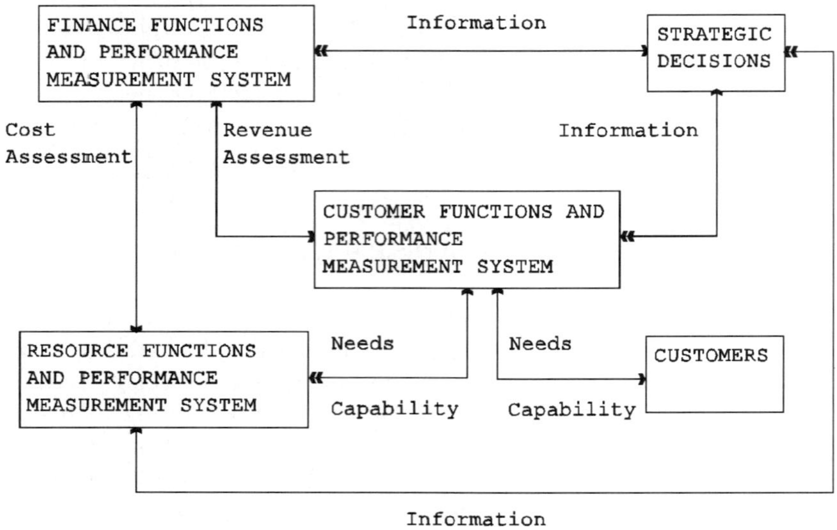

Fig. 3.1 An integrated performance measurement system framework.

assessing the financial impact of resource usage based on measures related to its strategic objectives. The benefits of integrated performance measurement systems are that they provide organizational linkages between:

- Strategies, objectives, policies, tactical plans and operational plans;
- Resource capability and customer needs;
- Financial requirements and market opportunities;
- Financial requirements and resource usage.

An integrated performance measurement system framework is illustrated in Fig. 3.1. The customer-based function (sales, marketing and field service) via market information recognizes the potential for increasing revenues by changing the current product offering to match customer needs and expectations. An assessment of the customer-based function's current capability to fulfill these needs is made through the customer-based performance measurement system. The market information is shared with the resource-based function (design, purchasing, production, warehousing, transportation and support functions) to compare the customer's needs to current resource capability through the resource-based performance measurement system. In addition, the finance-based function (finance and accounting) is given the market information to provide an assessment of future revenues generated from the new product offering, and the cost of any additional investments in resources required to produce the new product offering through the finance-based performance measurement system. Information on the results of the analyses conducted by the customer-, resource- and finance-based functions are forwarded to top management to support the strategic assessment of fulfilling the customer needs. Finally, the results of the strategic decision concerning the customer need are forwarded to the three functional areas.

Integrated performance measurement systems span between organizational levels and across functional areas to provide relevant information of strategic importance. These systems can therefore act as a decision support system (DSS), supporting decision makers in such areas as strategic planning, tactical planning and management control, and operational planning and control.

3.3 INTEGRATING PERFORMANCE MEASUREMENT SYSTEMS

Performance measurement systems comprise three characteristics: performance criteria, performance measures and performance standards (Crawford, Cox and Blackstone, 1988). Performance criteria are the relative elements used to evaluate performance (product yields, vendor delivery performance etc.). The actual values of the performance criteria over some specified time frame are the performance measures. Performance standards are the accepted levels of performance for each criterion. Traditionally, manufacturing performance standards were based on work measurement studies conducted by industrial engineers. Today, companies are setting standards based upon customer requirements and the competitive environment. Performance standards are no longer static and are constantly changing as firms continually improve their operations.

The hierarchical and functional management structures used by most firms create the need for performance measurement systems which monitor the firm's objectives at each organizational level and within each function. Mechanisms are also needed to provide vertical and horizontal linkages between the systems on specific strategic issues. These linkages may be accomplished in a variety of ways:

1. formal cross-functional interactions within organizational levels;
2. formal cross-functional interactions between organizational levels;
3. informal performance measurement information exchanges on strategic performance criteria through manual or electronic means.

A key to creating informal organizational linkages is a corporate culture which promotes teamwork. Without a team-oriented culture, the successful formation of vertical and horizontal linkages is not possible. Once established, such linkages result in improved performance by functional departments, more consistent decision making across organizational levels and better support on those activities necessary for achieving competitive advantages (Crawford, Cox and Blackstone, 1988, p. 2).

3.4 MANUFACTURING DECISION SUPPORT SYSTEMS

Davis and Hamilton (1993) state that since each business function (e.g. engineering, manufacturing and distribution) has unique information processing needs, computer applications and databases designed for a specific function are necessary. However, these systems must be designed within the context of the firm's management information system (MIS). The MIS must be capable of supporting the firm's strategic, tactical and operational planning activities. Also, the MIS must support management and operational control decisions and provide transaction processing and inquiry capabilities.

When information systems are used to assist management in problem solving, the term decision support system (DSS) is used to describe the human–machine interplay in reaching decisions (Hussain and Hussain, 1981). The systems are not intended to replace human decision makers, but to support human judgement (Davis and Hamilton, 1993). A DSS provides the link between the firm's strategic and tactical planning processes, and supports decision making for managerial and operational control.

A manufacturing decision support system (MDSS) must provide information to the manufacturing function which promotes decisions supportive of the firm's overall strategy.

Critical MDSS information includes:

1. product and process specifications;
2. customer order requirements;
3. production schedule timing and quantities;
4. material availability from vendors and preceding operations;
5. capacity requirements and capabilities of the existing workforce, equipment and facilities.

This information must be collected and transferred from other functional areas to support strategic manufacturing decisions on the transformation of operational inputs (human, machine, materials etc.) into finished products and customer orders. The information is also required to manage the flow of materials through manufacturing processes. However, the information consolidated and transmitted through the MDSS is needed not only to effectively coordinate transformation steps and material flows, but also to provide performance feedback for making continual operational improvements (Hayes, Wheelwright and Clark, 1988). Integrated performance measurement systems are MDSSs capable of supplying decision makers with the necessary information for ensuring that the manufacturing function is involved only in those activities which support the firm's strategy.

3.5 CASE EXAMPLE: THE HILL'S® PET PRODUCTS DIVISION OF THE COLGATE–PALMOLIVE COMPANY

3.5.1 Introduction

The Hill's® Pet Product Division of Colgate (located in Topeka, Kansas) is the industry leader in the production of specialized, veterinary-grade pet food products for the treatment of nutrition and dietary deficiencies in animals. The products are manufactured in two forms. Hill's® Pet Products' 'dry side' operation produces a pelletized product composed of various grains, dry poultry meals, vitamins and minerals. Dry finished product is packaged in 10, 20 and 40 lb (4.5, 9.1 and 18.2 kg) bags. The firm's 'can side' operation produces a 'wet' product composed of meat by-products, grain, vitamins and minerals. It is packaged in 6 and 15.5 oz (168 and 434 g) cans. There are two product types. Hill's® manufactures a 'science diet®' which provides life-cycle nutrition for dogs and cats from puppy and kitten to mature animal. The firm also produces science diets® which are considered 'lite products', designed for weight control and reduction in dogs and cats. Science diets® are sold through veterinarians and pet shops. The other product type produced by Hill's® Pet Products is a 'prescription diet®.' Prescription diets® are not medicated but are nutritionally balanced to help sick and diseased animals. The firm sells prescription diets® which aid in curing kidney, bladder, heart and other ailments in dogs and cats.

Hill's® customer base is 100% make-to-stock. The manufacturing processes are 100% continuous (high volume flow). There are three manufacturing facilities currently producing specialized pet food products at the following locations: Los Angeles, California; Bowling Green, Kentucky; and Topeka, Kansas. A new facility located in Richmond, Indiana, began operations in 1991. The Topeka plant has a total employment of 307, which includes 235 hourly and 72 salaried employees. Of the 235 hourly workers, 90 are classified as direct and 217 as indirect employees. The plant is a three-shift, eight-hour operation and produces 45 pet food formulations. There are 1500 active part numbers in the inventory system and approximately three levels in the typical bill of material for each product. Hill's® Pet Products is currently involved in implementing material requirement planning (MRP) and total quality management (referred to as 'total quality excellence' or TQE) division-wide, along with JIT and focused factory concepts for selected facilities, to effectively plan, monitor and control its manufacturing processes.

3.5.2 Strategic objectives

(a) Corporate objectives

The Colgate–Palmolive Company has a corporate objective of overall sales and profitability growth through the 1990s, with Hill's® playing a major role. Throughout the 1980s, Hill's® Pet Products represented approximately 10% of Colgate's sales volume, but contributed 25% of the corporate profits, along with substantial cash flow. Thus, Colgate's strategic objectives relative to Hill's® is to enhance the business so that revenue and profit growth can be sustained through the 1990s and beyond. A key to Colgate's growth objectives for Hill's® is the achievement of total quality excellence (i.e. total quality management) in all of its operations. To support this corporate initiative, Colgate has recognized Hill's® as a primary strategic business unit in its portfolio. The company is investing capital in a new facility, and in process control instrumentation for its existing facilities. Colgate is also encouraging Hill's® to develop business and manufacturing strategies which are consistent with corporate objectives.

(b) Business level objectives

Hill's® Pet Products has a mission to continue to be the pre-eminent supplier of small animal dietary management products in the world. The firm has defined five strategic objectives necessary to support overall corporate and business level strategies:

1. quality;
2. dependability (in servicing customer orders and equipment operation);
3. flexibility (product and process);
4. cost;
5. product development (i.e. product innovation).

Each plant has a manufacturing strategy which focuses on various combinations of the five strategic objectives. Thus, each facility is measured on its performance, based upon its manufacturing objectives. In this way, direct competition among the manufacturing operations is avoided since each plant has *different standards* for the same measures, based upon the strategic focus of the facility.

Hill's® Pet Products has also established specific order-qualifying criteria which apply to all manufacturing operations. These objectives are listed below.

1. **Superior product quality** – continue to produce the highest quality pet food on the market; the following are key elements to achieving this objective.
   (a) Remove as much variation from the process as possible to ensure product consistency.
   (b) Use certified vendors to ensure that the highest quality raw materials are used in the product.
   (c) Train employees on the importance of quality and in the techniques of total quality excellence.
2. **Maximum asset effectiveness** – use all resources to their maximum effectiveness; assets must be managed to allow the operation to support the capacity demands of the market.

(c) Manufacturing objectives

Hill's® Pet Products has developed strategic objectives in five areas for creating competitive advantages in the marketplace:

1. process flexibility
2. delivery and equipment flexibility
3. manufacturing cost
4. new product development
5. quality improvement.

Each manufacturing facility will make a unique contribution to the strategic success of the firm, based upon specific objectives established for the operation. Listed below are the manufacturing objectives for the four plants.

- **Richmond** – primary focus is on **flexibility**; the facility is designed to be a flexible plant and operates under the just-in-time (JIT) philosophy. The plant also focuses on the development of the most appropriate leading-edge technology for the business. Socio-technical concepts will also be explored in Richmond (i.e. development of the most effective work designs, training programs and continuous improvement efforts for the firm, given its technology).
- **Los Angeles** – primary focus is on **dependability**; it epitomizes the focused factory. The facility is small and landlocked, and produces a limited variety of finished product.
- **Bowling Green** – focus is on **dependability** and **cost**; it produces a simplified product mix at high volumes.
- **Topeka** – focus is **flexibility, dependability** and **product develop-**

ment. The dry side of the operation focuses on flexibility, and is currently running 18 different diet formulations. The can side focuses on dependability, and currently produces all the can products; since the Richmond plant opened, the can side has had long production runs with minimal changeovers. Part of the plant is being developed into a pilot facility to support product development activities.

All four facilities emphasize the production of high-quality products as part of their manufacturing objectives.

3.5.3 Performance measurement systems

An example of the performance measurement systems used at the plant and division levels is presented in Tables 3.1 and 3.2 respectively. The performance measurement systems used at the plant level are maintained by each department. The measures are collected functionally and consolidated by the plant accounting department. The plant accounting function provides plant and division management with periodic performance summary reports in the form of tables, charts and graphs. The summary reports are used to monitor trends and to develop improvement strategies. The division performance measurement system is based on financial and non-financial information received from the manufacturing facilities. Division management evaluates the manufacturing operations on the same criteria, measures and standards used by plant management. The primary differences between these systems are the level of detail and reporting frequency associated with each measure. The plant performance measurement system provides greater detail on each measure than the division system. Also, the performance measure reporting frequency is higher in the plant system than the division level system. Finally, the division system uses different performance standards on the same measure for each facility, based on its manufacturing objectives and focus.

3.5.4 Performance measurement system linkages

An example of how performance measurement information is transferred across functions and between levels to support tactical and strategic decision making is illustrated in Fig. 3.2. The Topeka plant has identified process flexibility as a key order-winning criteria for competitive success in its dry-side operations. Based on market needs, the operation has established a changeover performance standard of no more than 74 minutes. The daily average changeover time is measured and collected for each production department. At the plant level, a weekly changeover report is issued to all department managers. Information from the weekly reports is consolidated into a monthly performance report and is issued to

Table 3.1 Example of Hill's® Pet Products plant performance measurement system

Department(s)	*Performance criteria*	*Performance measures*	*Performance standards*
Plant staff/ accounting	Quality	Monthly scrap cost	Financial plan
		Monthly rework cost	Financial plan
		Weekly process yield (%)	84 (dry); 97 (can)
		Monthly process quality (%)	83 (dry); 95 (can)
		Quarterly AIB score	855
	Dependability	Monthly unscheduled downtime (%)	11 (dry); 16.5 (can)
		Monthly case fill rate (%)	95
	Flexibility	Monthly average changeover time (minutes)	74 (dry); 76.5 (can)
	Cost	Monthly total manufacturing cost improvement (%)	100% (no variances between actual and standard manufacturing cost)
	Safety	Monthly lost time frequency	3.7
		Monthly serious injury rate	10.1
		Monthly total incident rate	20.0

division managers. Plant management and staff personnel use the changeover information to make tactical and control decisions on plant flexibility performance. The division managers use this information to assess if the plant is achieving its flexibility objectives, and to determine if the plant's flexibility capability is supporting company sales and growth objectives.

Performance measurement system linkages between Colgate–Palmolive

Table 3.2 Example of Hill's® Pet Products division performance measurement system

Organizational level	*Performance criteria*	*Performance measures*	*Performance standards*
Division	Asset usage	Monthly scheduled downtime percentage	Division target for each plant
		Monthly unscheduled downtime percentage	Division target for each plant
		Monthly scheduled maintenance (%)	Division target for each plant
		Monthly unscheduled maintenance (%)	Division target for each plant
		Monthly machine usage (%)	Division target for each plant
		Monthly plant run time (%)	Division target for each plant
	Quality	Monthly scrap cost	Financial plan
		Monthly rework cost	Financial plan
		Monthly process yield (%)	Division targets for each plant
		Monthly process quality (%)	Division targets for each plant
		Quarterly AIB score	Division target for each plant
	Dependability	Monthly case fill rate (%)	Division target for each plant
	Flexibility	Monthly average changeover time	Division targets for each plant
	Cost	Monthly total manufacturing cost improvement ((%))	100(%) (no variances between actual and standard mfg. cost)
		Monthly production cost per pound	Division targets for each plant

and Hill's® Pet Products are primarily accomplished via periodic financial reports. Hill's® is currently measured by Colgate corporate management against its profit plan and volume growth objectives. The division is using asset usage measures as a link to the profitability and volume measures assessed by corporate staff. Hill's® Pet Products division management believes that effective asset usage has a direct impact on corporate profitability and growth objectives, thus providing linkages to corporate measures. The objectives will be accomplished through enhanced asset effectiveness via improved plant uptimes, plant yield improvements, reductions in plant unscheduled downtime, more effective preventive maintenance etc. At the last Colgate profit review, Hill's® division management shared asset effectiveness results with the corporate staff for the first time. Division management will continue to inform Colgate management on asset usage improvements in this manner.

Performance measurement system linkages between division management and the plants are accomplished primarily through individual monthly performance reports which present performance results in the following areas:

1. quality
2. dependability
3. flexibility
4. cost
5. safety.

The reports are sent from the plant managers to the director of manufacturing. The director of manufacturing distributes the reports throughout the division manufacturing function. Currently the plant performance results are shared informally with other division functions. Plans are underway to develop formal mechanisms for sharing manufacturing performance results with all division functions. The director of manufacturing technology was developing an executive reporting system which would further enhance performance measurement system linkages between division and plant management. Division and plant level performance measurement linkages are also accomplished through quarterly performance presentations made to functional vice-presidents by plant management personnel. The presentations focus on current plant performance and action plans for future improvement.

Several vehicles are used within the plants to link departmental performance measurement systems and information on plant organizational performance to plant personnel. Mechanisms used for plant-wide performance measurement linkages are illustrated below.

- **Daily department operations report** – generated by can and dry department supervisory staffs daily, and indicates the weight of product extruded/processed and weight packed per shift. It also provides daily totals and a running monthly total, used to monitor departmental performance against commitments. It is distributed to the can and dry department managers.
- **Daily morning recap meeting** – 8:00 a.m. plant management staff meeting to discuss plant activities which have occurred in the last 24 hours; also to discuss targets for the current day and important events scheduled for the next two to three days.

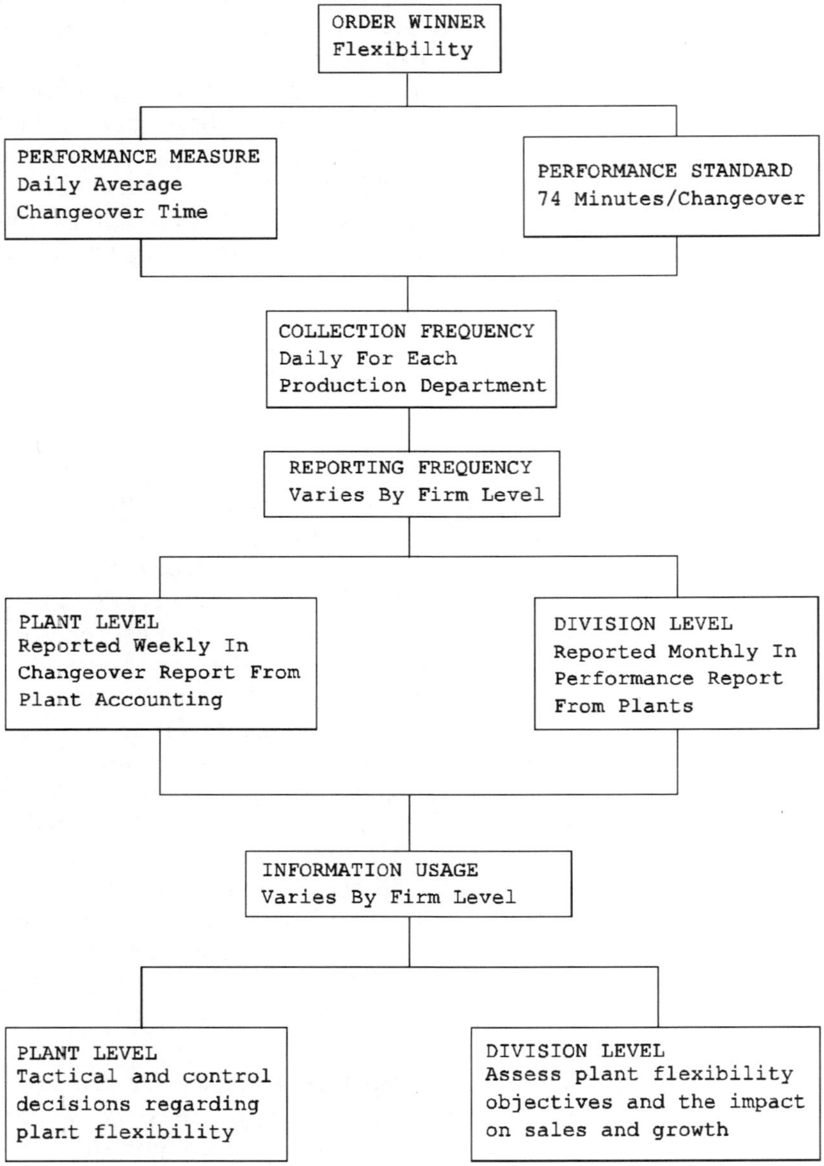

Fig. 3.2 An example of performance measurement information usage to support tactical and strategic decisions.

- **Weekly yield report** – contains yield performance information for the can and dry operations; it is distributed to plant management staff.
- **Weekly downtime and average changeover time report** – contains downtime and changeover information for can and dry operations; it is distributed to plant management staff.
- **Weekly work order recap report** – describes work order completions by department; it is distributed to division engineering group and plant management staff.
- **Weekly department meetings** – internal meetings by each department to assess its effectiveness in supporting plant objectives and to develop action steps for improvement.
- **Weekly plant staff meetings** – to examine plant performance for the week; analyze key performance measures in each category (quality, dependability, flexibility etc.); develop action plans to enhance performance.
- **Monthly performance report** – highlights plant performance in the areas of quality, dependability, flexibility, cost and safety; it is distributed to plant management staff and the director of manufacturing.
- **Quarterly performance summary** – key performance results in the above-mentioned categories, summarized in graphs, tables and charts. It is posted in the manufacturing area, and provides feedback on plant performance to production workers.

In addition, the plants have a 'Quality pays' bonus system based on the number of customer complaints per quarter and on American Institute of Bakers (AIB) quality scores. A bonus is paid quarterly to all employees according to the results of these measures. AIB scores are announced plant-wide each quarter via the intercom system. Thus, monetary incentives have been installed to support plant quality efforts. The techniques used by Hill's® Pet Products to achieve performance measurement system linkages between organizational levels and departments congruent with corporate, business level and manufacturing objectives have been designed to help the firm to maintain its leadership position in the specialized pet food industry.

3.6 CONCLUSIONS

The Hill's® Pet Products case illustrates two key concepts essential to using integrated performance measurement systems as an MDSS providing linkages between systems and functions on strategic objectives.

1. **Integrated performance measurement systems must contain measures which focus organizational resources on achieving strategic objectives**. Hill's® defines dependability in servicing customer orders

as a strategic objective. Therefore, the firm actively manages its organizational resources in a manner which maximizes asset effectiveness. Assets are managed to allow the firm's operations to support the capacity demands of its served markets. Key dependability metrics used by the firm are case fill rate, defined as the monthly percentage of customer orders filled (an indicator of dependability to the customer) and the percentage of unscheduled downtime (an indicator of equipment dependability). These metrics are used to focus organization resources on improvement efforts leading to better support of customer delivery expectations.

2. **Integrated performance measurement systems must be capable of monitoring the impact of resource usage on strategic performance.** The impact of improved resource usage on the firm's financial performance is monitored via the total manufacturing cost improvement index. This index is defined as the monthly total manufacturing cost divided by the monthly total standard cost, multiplied by 100%. The index allows Hill's® to examine the financial impact of plant improvement efforts (i.e. improvements in resource deployment). The plants are also beginning to focus on tracking those activities which 'drive' the use of organizational resources. Thus the primary emphasis has shifted from a pure cost focus to an activity focus using measures which show the impact of resource usage on its strategic objectives. The use of integrated performance measurement systems as an MDSS can provide a firm with the necessary linkages for achieving manufacturing excellence.

REFERENCES

Crawford, K.M., Cox, J.F. and Blackstone, J.H. Jr (1988) *Performance Measurement Systems and the JIT Philosophy: Principles and Cases*, APICS, Falls Church, VA.

Davis, G.B. and Hamilton, S. (1993) *Managing Information: How Information Systems Impact Organizational Strategy*, Business One Irwin, Homewood, IL.

Hayes, R.H. Wheelwright S.C. and Clark, K.B. (1988) *Dynamic Manufacturing: Creating The Learning Organization*, The Free Press, New York.

Hill, T. (1989) *Manufacturing Strategy: Text and Cases*, Irwin, Boston, MA.

Hussain, D. and Hussain, K.M. (1981) *Information Processing Systems For Management*, Richard D. Irwin, Homewood, IL.

Johnson, H.T and Kaplan, R.S. (1987) *Relevance Lost: The Rise and Fall of Management Accounting*, Harvard Business School Press, Boston, MA.

Lockamy, A. III and Cox, J.F. III (1994) *Reengineering Performance Measurement: How to Align Systems to Improve Processes, Products, and Profits*, Irwin Professional Publishing/APICS, Burr Ridge, IL.

CHAPTER FOUR

Intelligent decision support for quality function deployment

S. Daniel Bird

4.1 INTRODUCTION

The role of decision support and expert systems in manufacturing has been slowly growing over the past few years. One reason for the increased usage of such systems is that many production problems are far too complex for traditional optimization methods (Philip and Schultz, 1990). The application of support technologies to manufacturing problems, however, has primarily remained in the more *reactive* domains of diagnosis, monitoring and control. The view taken in this chapter is that intelligent support systems should be developed to take a more *proactive* role in modern industrial organizations.

This proactive role can be achieved, in part, by applying support systems technology to the processes of quality management for the construction of (semi-)intelligent quality management decision aids. Because quality management has high potential payoff, it stands to reason that decision aids which support these activities have high potential payoff as well. It has been speculated that support systems, and particularly those with some degree of 'intelligence' like expert systems and intelligent decision support systems, will become 'more important tools as U.S. industry moves from limited manufacturing quality control to more comprehensive product development quality control in which many more people, including many not in quality departments, work on [quality control] teams' (Braun, 1990, pp. 72–73). Support system applications may not only be instrumental in, but also a prerequisite for the wholesale adoption of quality management techniques such as quality function deployment (QFD) and quality planning (QP) (Bird, 1992; Braun, 1990). Such developments can also help bridge the gap between the information processing and production functions of the adaptive, computer-integrated manufacturing enterprise.

4.2 ACTIVE VERSUS PASSIVE QUALITY

Although already having a strong foundation, published accounts of manufacturing decision support and expert systems are increasing in number and variety. The growth in the application of these technologies to specific manufacturing needs is well-documented (e.g. *Computers in Industry*, 1990; Oliff, 1988). A sampling of recent papers detailing their use in manufacturing shows applications in job shop scheduling (Randhawa and McDowell, 1990), equipment diagnosis (Bajpai, 1988), equipment acquisition and evaluation (Padmanabhan, 1989), rotational machining tolerances (Mannivannan, Lehtihet and Egbelu, 1989), surface defect, inspection (ElMaraghy and Bullis, 1989), scheduling materials handling devices (Thesen and Lei, 1990), process planning (Chitta, Shankar and Jain, 1990), and design (Floss and Talavage, 1990; Heragu and Kusiak, 1987) among innumerable others.

More general reviews within application areas include Majstorović's (1990) survey of nearly 90 expert systems for diagnosis and maintenance and Alting and Zhang's (1989) survey of more than 150 computer-aided process planning expert systems. Bird (1992) provides an even broader view of expert systems in the application domains noted above (diagnosis, monitoring and control), as well as the domains of debugging, design, interpretation, instruction, planning, prediction and repair. That paper also provides a formalization of quality management support systems, although it only sketches two limited-functionality, experimental prototype architectures. Other 'quality'-oriented papers which promote the use of information management and support system technologies to address the demands of the quality movement include a data interpretation and diagnosis expert system (Zumegen, 1989), an expert system for integrated circuit wafer fabrication (Sitte and Harrison, 1991) and managerial quality information system considerations (Lin, 1991). In short, the number of published accounts of systems which support quality control or quality management techniques are still comparatively few, but increasing.

One reason for this relatively slow growth is that manufacturing decision support and expert systems have been developed without the specific intent to address quality problems. Claims of improvement in product or process quality from their usage, however, are common. This type of quality improvement has been argued elsewhere as essentially a windfall (Bird, 1992), a by-product of support system technology usage. Although quality improvement is often not an initial consideration in their development, support systems 'invariably help enhance quality and the automation of routine procedures that, in turn, reflect on the quality issues that are associated with every job in industry' (Braun, 1990, p. 74). This type of quality improvement, achieved incidentally from support systems usage, is thus passive.

As suggested, the full realization of benefits from quality management methods may ultimately be linked to the effective application of information and support system technologies to facilitate their use. That support will necessarily have both 'proactive' and 'reactive' aspects. It is the proactive, however, which has the potential to transform quality improvement into an active outcome from support system usage, instead of merely a passive one.

4.3 QUALITY FUNCTION DEPLOYMENT SUPPORT SYSTEM

Quality as a continuous, feedforward process is one of the fundamental principles of the modern repertoire of quality management methods such as the quality function deployment (Akao, 1990) and quality planning (Juran, 1988) techniques. Of the many emerging quality management techniques, QFD is clearly among the most widely embraced. A brief overview of the quality function deployment methodology is first presented in this section, along with its impact on product and process quality. A preliminary architecture for the design of an intelligent support system for the basic QFD method is then outlined.

4.3.1 The QFD Method

QFD is a set of quality management principles supported by a structured method for converting customer demands into 'quality characteristics' and then developing a product/process design by systematically deploying the relationships between customer demands and those characteristics (Akao, 1990). The focus of QFD is on analysis and documentation, with the method breaking down into numerous analytical steps which are documented as a set of charts. A widely accepted structured method for applying the QFD philosophy as well as for creating its deliverables has been developed by Akao (1990).

QFD is generally described as a four-phase process:

- product planning
- part deployment
- process planning
- production planning.

A number of deliverables, again generally in the form of charts, are produced in each of these phases. In fact, QFD relies heavily on charts not only to document product and process information, but also to facilitate analysis. The high-level inter-relationships among these phases are

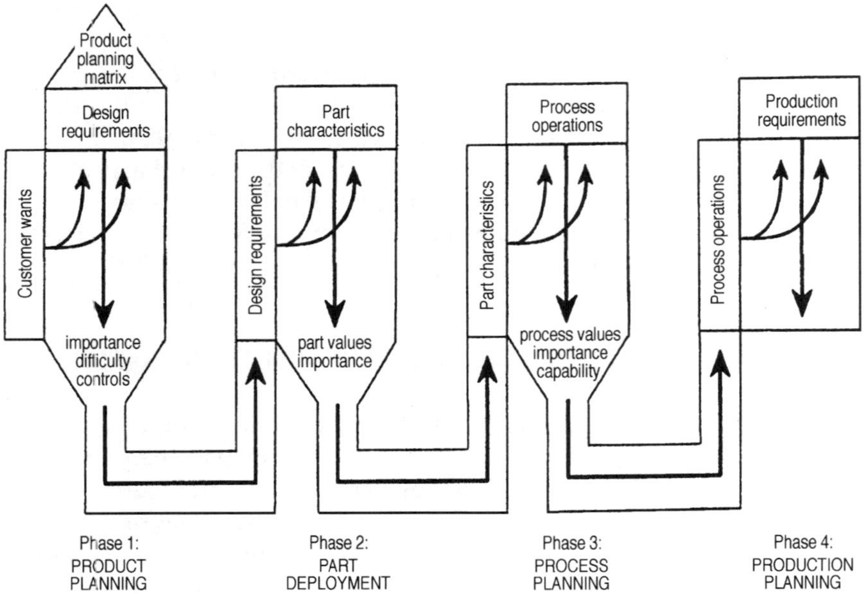

Fig. 4.1 Main phases of quality function deployment and their inter-relationships.

graphically illustrated in Fig. 4.1, along with the major chart (or matrix) of each phase.

As this figure illustrates, the basic tool of QFD, the **product planning matrix**, drives the QFD technique. This matrix 'contains the most critical information you need regarding your relationship to consumers and your competitive position in the marketplace' (American Supplier Institute, 1989, p. 4-3). Stated customer needs, translated into design requirements, are documented in this chart in terms of their relationships, correlations, perceived measures of organizational difficulty, competitive engineering assessments, services repairs and costs, controls, and technical importance.

These items serve as inputs to the **part deployment matrix** which facilitates the translation of product design characteristics into assemblies and parts, part characteristics, as well as objective target values for part design requirements. These are similarly supplemented with critical part characteristic values and perceived measures of importance. The information documented in this chart then feeds into the **process planning matrix** which details component and material flows, process elements, and critical process parameters. Also associated with this chart are assessments of process capability, the values of critical process parameters, and evaluations of their importance.

Finally, the **production planning matrix** provides a medium for the translation of material and component flows, as well as process elements, parameters and values into assessments of operational evaluation and planning requirements. Documented measures generally include those for operation difficulty and frequency, control chart usage, training and job instruction, preventative maintenance schedules and process cycle times, among others.

Although it has taken time to filter into American industry, QFD has been successfully applied to everything from developing software to strategic planning. Its primary usage, however, remains in the manufacturing sector where it is applied to the design and manufacture of physical goods. In this setting, innumerable benefits have been realized: shorter time-to-market, increased quality and customer satisfaction and enhanced organizational learning (O'Connor *et al.*, 1992).

4.3.2 QFD support

'The QFD process leads the participants through a detailed thought process, pictorially documenting their approach. The *graphic* and *integrated* thinking that results, leads to the preservation of technical knowledge, minimizing the *knowledge loss* from retirements or other organizational moves. It also helps transfer knowledge to new employees, starting them *higher on the learning curve*' (American Supplier Institute, 1989, p. 2–8, emphasis in original). QFD is not, however, an easy methodology to learn or to apply. In fact, it has many documented limitations: little support for its numerous required translations, time-consuming and skilled-labor intensive, limited provision of decision rationalization and little support for the systematic comparison of alternatives (O'Connor *et al.*, 1992).

Despite its limitations, the graphic and integrated nature of the QFD process provides a formal approach for dealing with the complexities of product and process development. Its hierarchical structure and conditional evaluation of product and process parameters have emerged as a highly functional, 'natural' architecture (Hollnagel, 1987a,b) for the product/process development problem. That is, the QFD methodology takes advantage of the fact that decision-maker-defined structure of data and knowledge is closely related to its usage.

This same relationship ties the techniques of QFD to knowledge processing technologies – technologies which have the potential to overcome many QFD limitations. In particular, the structure of the matrices fits the rule-based knowledge-representation paradigm for expert systems quite well: both use a basic premise/conclusion structure. Several of the noted QFD benefits, in fact, echo those realized

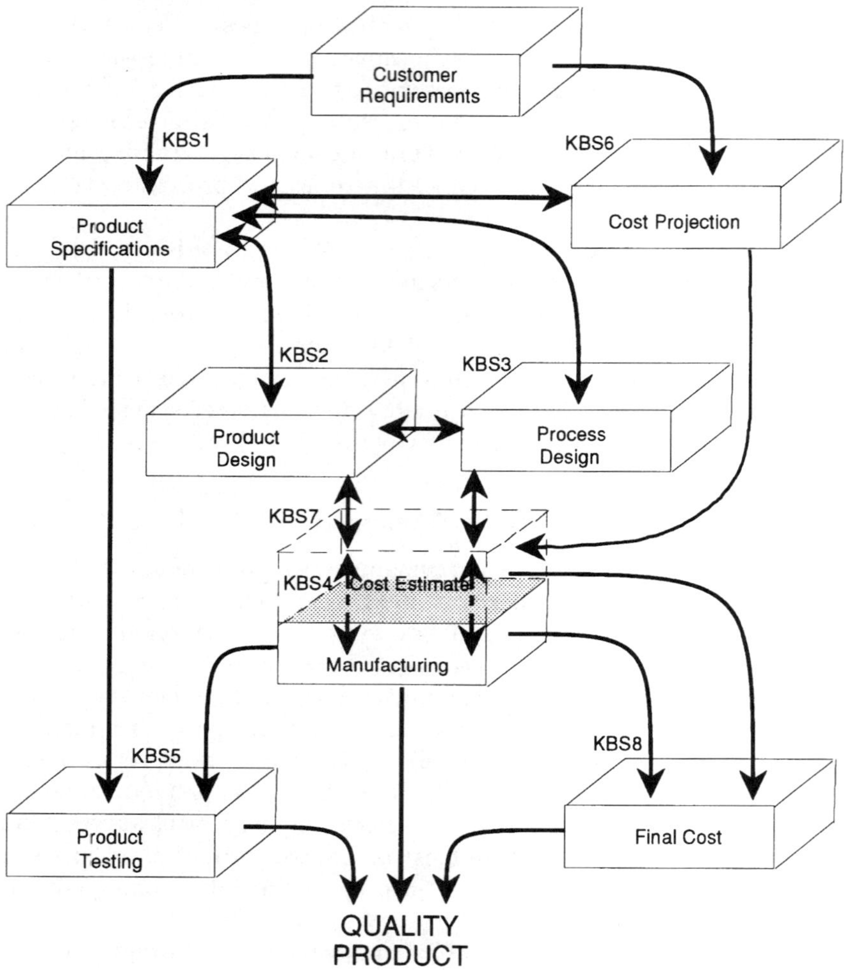

Fig. 4.2 QFD activities and knowledge base support.

from the use of decision support and expert systems. Because the techniques of QFD and (formalized) expert problem solving employ a common structure for data and knowledge representation – a structure that has functionally emerged and that is preferred by the decision maker – it reasons that (intelligent) decision support system (IDSS) can be designed to facilitate the problem-solving activities of QFD.

'From the point of view of actually designing and implementing IDSSs, the important point is that an IDSS is able to amplify the strong features of the user's thinking and supplement the weak points. This relates not only to the way the system reasons, but also to the way knowledge is represented ...' (Hollnagel, 1987b, p. 744). Noting the matrix-rule 'fit' and that a network of systems can be used to unite the various elements of the total QFD methodology, Braun (1990) proposes a network of knowledge-base systems to support QFD. A knowledge-base model for QFD support, adapted from Braun, is illustrated in *Fig. 4.2.*

Starting with **customer requirements** (or wants), as in the figure, QFD participants translate those requirements into **product specifications**. Knowledge-base techniques facilitate this (via KBS1) as well as all subsequent translations. At this stage, requirements are also beneficially translated into a **cost projection** for the product (via KBS6). The formal technical product specification is then simultaneously translated into a **product design** (via KBS2) and a **process design** (via KBS3), as well as a set of **product testing** criteria (via KBS5) and a revised cost projection (again using KBS6). This 'simultaneous' or 'concurrent' design of product and process helps ensure the manufacturability and economic feasibility of the product. The initial specification of acceptance criteria also ensures the objectivity of any quality control evaluation.

Product and process designs are then used to develop an accurate *cost estimate* for the product and to establish production economics (via KBS7). This cost estimate must, of course, be reconciled against the initial cost projection and managerial decisions must be made to abandon, continue or re-engineer the product and process, as well as to evaluate any final **manufacturing** considerations or constraints (via KBS4).

Given that the decision is made to proceed with manufacturing, the final steps are **product testing** (again using KBS5) and determining the product's **final cost** (via KBS8). Product testing attempts to uncover flaws or inefficiencies in both the product and process, not just provide final product inspection. Again, the production cost must be reconciled against previous cost estimates. Finally, it should be noted that the basic QFD method is supported by knowledge bases KBS1–KBS5 in this model. The three additional knowledge bases, KBS6–KBS8, are cost assessment supplements of indisputable importance.

4.3.3 Support system development

The structure of the QFD method, organization of its procedures and deliverables, and the degree of coupling between the deliverables and procedures for their creation offer direction in developing a support system architecture. From this, Bird (1992) provides an object-based architectural description of how **rules** and **procedures** can be employed to apply the steps of the QFD method as well as how **objects** can be used to both store the structure of the various deliverables and provide type hierarchies and inheritance. In that architecture, **states** aggregate various processing and inferential elements such as rules and procedures. The various data – grouped into **classes** that describe the structure and content of data attributes – are then class **instances**, each of which represents a segment of some QFD chart. O'Connor *et al.* (1992) provide a description of another QFD matrix-driven system, the proprietary SeeQFD™ software. Although primarily a description of system capability, their paper relates similar functional system characteristics: heuristic-based aiding, support for decomposition and integration, the preservation of expertise and other functionality.

As these two papers illustrate, the construction of a decision support system for the QFD technique is an undertaking of immense complexity. Consider, for instance, *only* the structure and content of the knowledge base for process design (KBS3), or more generally process planning. Process planning is the formulation of a collection of tasks for translating component designs into a machining specification which converts a piece or part from initial to final forms. The resulting step-by-step specification of machining tasks, the process plan, yields not only a procedure to create a finished part, but also provides the foundation of production control. 'The process plan forms the basis for cost accounting and other performance evaluation practices, developing detailed schedules, data collection and monitoring, equipment engineering, and organizing the shop floor (facilities design)' (Bhaskaran, 1990, p. 1528).

Two basic computer-assisted process-planning techniques have been defined: variant process planning and generative process planning (Alting and Zhang, 1989). The variant approach generally uses group technology to classify and code each part and a database to store process plans. Variant-based process plans for new parts are created by first classifying and coding the part and then using the process plan of a similar existing part as a starting point. The generative approach differs in that instead of storing process plans, the software captures part, material, machine, tool and fixture information as well as the logic or knowledge required for generating a process plan. From the component design and the knowledge stored in the system, a complete process plan is generated for each new part.

Accordingly, a human or automated process planner takes a part design – usually a CAD model – and from production environment knowledge, 'experience' and part design specifications, develops the steps in, and sequence of, production activities to produce that part. Knowledge of current manufacturing capabilities, design interpretation and planning logic are thus fundamental to the activities of process planning. The phases of process plan, its logic flow and knowledge base content for variant-based process planning support are shown in Fig. 4.3.

As illustrated in the figure, performing process planning starts with knowledge of the coding and classification techniques of group technology as well as analytic and knowledge-based techniques for interpreting part topology and geometry. It then demands the knowledge

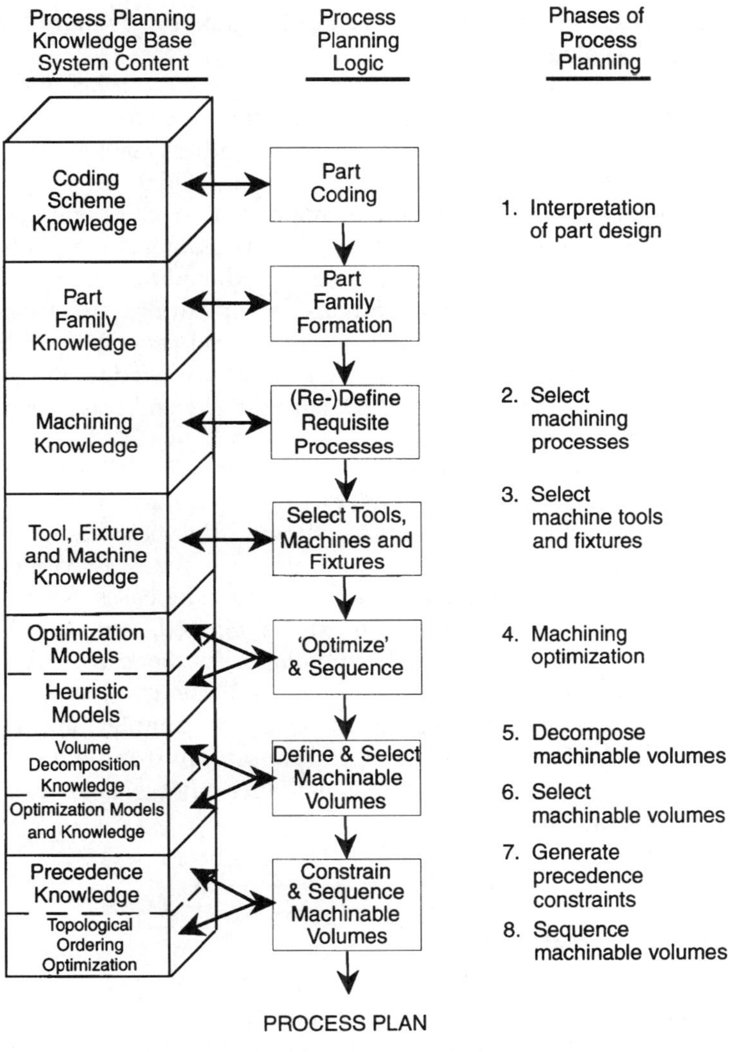

Fig. 4.3 Simplified variant process planning: content of its requisite knowledge base (KBS3), logic flow and general phases.

of machining processes which yield identified part characteristics as well as that related to the properties and applicability of machines, tools and fixtures. Machining must be 'optimized', requiring the use of both optimization and heuristic models. Knowledge-based volume decomposition techniques must then be applied, again requiring the use of optimization and heuristic models for the selection of machinable volumes. Knowledge of machining precedence relations and constraints must then be applied and the machining volumes must be optimized, requiring further optimization models. Finally, the process plan must be generated and evaluated.

All of this, it should be noted, is typically done without knowledge of shop-floor status. Hence, tool breakage, machine failure, maintenance activities, delays and priorities and many other factors not only complicate this (simplified) process, but often invalidate its output. As per QFD, however, the translation of material and component flows, process elements, parameters and values into assessments of operational evaluation and planning requirements (the production planning matrix, above) generally include considerations of operation difficulty and frequency, preventative maintenance schedules and process cycle times. Thus, the process planning 'system', as a set of knowledge and model bases, must interact with other QFD support system components to achieve the benefits of integration.

Finally, this figure intimates the scope and complexity of the development effort when attempting to provide computer support for quality management techniques. As shown, the 'knowledge base' required to support one QFD activity may be many separate, but interacting and interdependent knowledge and model bases. Knowledge and decision models, of course, must be elicited, structured and adequately represented, tested and maintained. As the knowledge base increases in size (i.e. amount of knowledge), the complexity of its expansion and maintenance compounds. Other complicating factors include normal changes in the production environment – new machines, tools, fixtures, and materials handling devices, changes in the floor layout and so on.

4.4 CONCLUSION

The use of decision support systems and expert systems in manufacturing is widely accepted and growing. A number of such systems have been developed and delivered with notable quality-related benefits. Most of these systems, however, assist a decision maker only in identifying problems in current products and production methods and thus were characterized as reactive quality tools. Because these systems provide only limited support for the proactive use of quality management techniques, they yield only modest success toward achieving 'total' quality.

To assist manufacturers in taking a proactive approach toward total quality management, methods such as quality function deployment and quality planning have been developed. As argued, QFD (and QP) is directly amenable to decision support and expert systems techniques. This chapter has attempted to illustrate not only the applicability of intelligent support systems to the QFD quality management method, but also the difficulty inherent in efforts to realize such support systems. The widespread usage and ultimate success of QFD and other quality management methods in manufacturing may ultimately depend on the successful development of this type of computer-facilitated support.

REFERENCES

Akao, Y. (1990) *Quality Function Deployment*, Productivity Press, Cambridge, MA.

Alting, L. and Zhang, H.C. (1989) Computer aided process planning: the state-of-the-art survey. *International Journal of Production Research*, **27**(4), 553–85.

American Supplier Institute (1989) *Quality Function Deployment: Three Day Workshop, Implementation Manual*, Ver. 3.4, Dearborn, MI.

Bajpai, A. (1988) An expert system model for general-purpose diagnostics of manufacturing equipment. *Manufacturing Review*, **1**(3), 180–87.

Bhaskaran, K. (1990) Process plan selection. *International Journal of Production Research*, **28**(8), 1527–39.

Bird, S.D. (1992) Object-oriented expert system architectures for manufacturing quality management. *Journal of Manufacturing Systems*, **11**(1), 50–60.

Braun, R.J. (1990) Turning computers into experts. *Quality Progress*, February, 71–75.

Chitta, A.K., Shankar, K. and Jain, V.K. (1990) A decision support system for process planning. *Computers in Industry*, **14**(4), 307–18.

Computers in Industry (1990) Special issue: *Intelligent Manufacturing Systems – IMS '89*, **15**(1/2).

ElMaraghy, H.A. and Bullis, D.J. (1989) Expert inspector of surface defects. *Computers in Industry*, **11**(4), 321–31.

Floss, P. and Talavage, J. (1990) A knowledge-based design assistant for intelligent manufacturing systems. *Journal of Manufacturing Systems*, **9**(2), 87–102.

Heragu, S.S. and Kusiak, A. (1987) Analysis of expert systems in manufacturing design. *IEEE Transactions on Systems, Man, and Cybernetics*, **17**(6), 898–912.

Hollnagel, E. (1987a) Information and reasoning in intelligent decision support systems. *International Journal of Man–Machine Studies*, **27**(5/6), 665–78.

Hollnagel, E. (1987b) Commentary: issues in knowledge-based decision support. *International Journal of Man–Machine Studies*, **27**(5/6), 743–51.

Juran, J.M. (1988) *Planning for Quality*, Free Press, New York.

Lin, B. (1991) Quality control information systems in manufacturing: considerations and concerns for management. *International Journal of Operations and Production Management*, **11**(1), 41–50.

Majstorović, V.D. (1990) Expert systems for diagnosis and maintenance: the state-of-the-art. *Computers in Industry*, **15**(1/2), 43–68.

Manivannan, S., Lehtihet, A. and Egbelu, P.J. (1989) A knowledge based system for the specification of manufacturing tolerances. *Journal of Manufacturing Systems*, **8**(2), 153–60.

O'Connor, L., Partridge, D., Seely, B., Guthmiller, W. and Lovette, K. (1992) The SeeQFD™ software: an environment for QFD. *SAE Technical Paper Series, No. 922117.*

Oliff, M.D. (ed.) (1988) *Expert Systems and Intelligent Manufacturing, Proceedings of the Second International Conference in Production Planning and Control*, North-Holland, Amsterdam.

Padmanabhan, S. (1989) A tandem expert support system as justification for a flexible manufacturing system. *Journal of Manufacturing Systems*, **8**(3), 195–205.

Philip, G.C. and Schultz, H.K. (1990) What's happening with expert systems. *AI Expert*, **5**(11), 57–59.

Randhawa, S.U. and McDowell, E.D. (1990) An investigation of the applicability of expert systems to job shop scheduling. *International Journal of Man–Machine Studies*, **32**(2), 203–13.

Sitte, R. and Harrison, H.B. (1991) Use of an expert system for quality improvement in flexible manufacturing of integrated circuits. *Computers in Industry*, **16**, 35–38.

Thesen, A. and Lei, L. (1990) An expert scheduling system for material handling hoists. *Journal of Manufacturing Systems*, **9**(3), 247–52.

Zumegen, P. (1989) Knowledge based quality information system for industrial application, in *Proceedings of The Second International Conference on Industrial and Engineering Applications of Artificial Intelligence and Expert Systems*, IEA/AIE-89, The University of Tennessee Space Institute, June, pp. 655–63.

CHAPTER FIVE

A knowledge-based decision support system for apparel enterprise evaluation

Sambasivan Narayanan and Sundaresan Jayaraman

5.1 INTRODUCTION

Until recently, determining the ability of a manufacturing enterprise to produce and deliver a commodity has largely been a highly subjective process rather than an informed decision-making process. The traditional approach relies heavily on the expertise, knowledge and experience of the evaluator. However, with the advancements in knowledge-based systems (KBS) and decision support systems (DSS), it is feasible to abstract and represent the expert knowledge and develop a knowledge-based approach. A knowledge-based decision support system (KBDSS) can determine various factors which affect the manufacturing capabilities and how these factors can be aggregated to arrive at a quantitative index for the enterprise. Such an approach will result in consistent and reliable evaluations. Moreover, such expert knowledge can widely be dispersed throughout an enterprise. In this chapter, we present the details of the design and implementation of a knowledge-based system, which can assist decision makers to evaluate apparel manufacturing enterprises.

5.1.1 Decision support for source selection

The practice of subcontracting is prevalent in many industries. The buying organization typically receives bids from several companies and selects one of them to carry out these operations as a subcontract. This process is known as **source selection**. Selecting the lowest bidder may appear to be beneficial at the time of awarding the contract, but it may not necessarily turn out to be the overall best value decision. This is because the total cost involved in the specific lowest bid contract may be higher than the initial bid, as a result of poor quality, or failure to fulfill the buyer's order on time. In order to get the best overall value for the buyer, there is a trade-off between the price quoted in the bid and the ability of the bidder to fulfill the contract efficiently. The objective of this knowledge-based system is to provide reliable decision support to quantify and evaluate the contractor's technological competence and ability to meet or exceed the performance requirements.

Deriving objective performance indexes from available subjective data and using them for capability evaluation is not uncommon. The Federal Aviation Administration (FAA) has derived performance indicators based on on-time arrivals and departures, lost baggage etc. to evaluate the capabilities of different carriers. But no such performance indicators exist for evaluating apparel enterprises. In this chapter, we outline the methodology we incorporated to arrive at reliable indicators for evaluating apparel manufacturers and the implementation of this methodology in a knowledge-based decision support system. Once the validity/applicability of such an approach to apparel manufacturing enterprises is determined, it can be extended to other domains such as food, metal parts and medicine.

5.1.2 Role of the apparel enterprise evaluation system in manufacturing enterprise architecture (MEA)

Computer-integrated manufacturing (CIM) involves integrating computers in various functions of an enterprise to produce the right product at the right time, right quality and right price (Jayaraman, 1990). Therefore, an important prerequisite for the implementation of CIM is the in-depth knowledge about every function in the enterprise. A complete and structured definition of the knowledge of all the fundamental functions of the manufacturing enterprise is known as the manufacturing enterprise architecture (MEA) (Jayaraman, 1989a). The source selection process is one such functional component of MEA and should be automated to the extent possible in order to achieve complete integration in an enterprise. Therefore, a framework encompassing the source selection process knowledge is an integral part of MEA. Such a framework for evaluating the capabilities of apparel enterprises is known as the apparel enterprise evaluation framework (AEEF).

5.1.3 Project initiative

The US Department of Defense (DoD) is the largest buyer of apparel items in the western world, procuring approximately $1 billion worth of apparel through contracting (DPSC, 1988). As part of the procurement process,

DoD is required to evaluate the manufacturing facilities of bidders on contracts. The old practice of using sealed bid procedures and awarding contracts to the lowest bidder is giving way to **best value procurement**. Bidder selection will be more effective if reliable quantitative methods for evaluating the contractor's potential could be developed. An approach based on acquiring the existing compiled knowledge of apparel manufacturing and source selection from contracting experts, and representing it in a structured knowledge framework, can aid the development of such methods and indicators. This informed knowledge-based framework can benefit not only DoD, but also the whole apparel industry in general, since subcontracting is very prevalent in the apparel industry. If a standard set of complex rules could be developed to act upon the knowledge of the bidders' technical capabilities, it can be used as a framework for evaluating them. Moreover, this approach has the potential to be extended to any type of manufacturing enterprise, by carrying out appropriate modifications in the knowledge-based framework.

5.2 PROJECT OBJECTIVE

The objective of this project has been to develop a system to assist evaluators in improving the quality of the decision-making process in source selection in apparel procurement. This objective has been achieved through the following steps:

1. Designing and developing a knowledge-based framework (the AEEF) to determine the major factors which affect the capabilities of an apparel enterprise and how each of these factors affects the overall possibility of getting a quality product at the right time from that enterprise;
2. Implementing this knowledge-based framework in a decision support system, which can be used by apparel buyers to evaluate the capabilities of their contractors' apparel manufacturing facilities;
3. Developing a front-end user interface to obtain the necessary information from the contractors.

Utility trouser manufacturing has been chosen as the domain for this research because it represents a significant segment of items procured by DoD (approximately 300 000 pairs per year). Once the system is developed for utility trousers, the framework can be augmented to include other apparel items.

The major effort in building the knowledge framework involves the transformation of the knowledge of measurable quantities obtained from the bidders (e.g. technology level of sewing machines, average experience of machinery operators, number of QC inspectors) into entities of higher levels of abstraction such as production capability, quality capability and financial capability. In the AEEF, the transformation of observable data into higher levels of knowledge is based on information from literature, the analysis of responses to a questionnaire to experts in the area of apparel manufacturing and contracting, and discussions with experts.

5.3 CURRENT PROCUREMENT AND SOURCE SELECTION PROCEDURES

As the first step in building this knowledge-based decision support system, the literature in the area of source selection has been reviewed. The US Defense Logistics Agency (DLA), the Defense Personal Support Center (DPSC) and DoD handbooks provide detailed descriptions of procurement and the formal source selection process with special reference to clothing and textiles contracts (Edwards, 1989; DPSC, 1987, 1988). DPSC also has a list of acceptable suppliers for use by DLA contractors in subcontracting (ASL, 1985). If DLA contractors procure raw materials from these acceptable suppliers, they are not required to perform raw material inspection and testing.

Lange and Heuermann performed an in-depth analysis of the army's contractor evaluation program (Lange and Heuermann, 1973). They concluded that past performance was the criterion used universally. In the context of vendor evaluation based on informal sources of information, they inferred that those efforts were at best, marginally effective. While reviewing the private industry's practices, they found that although **vendor evaluation** efforts are performed sometimes, **vendor rating** efforts are extremely rare. They also state that a few major companies had tried to establish formal, elaborate vendor rating systems, but abandoned these efforts, because they were found to be generally unworkable, unmanageable and often ineffective. The vendor rating systems were discontinued because the efforts required for maintaining the system as an effective management tool were not justified by the results achieved. In spite of these failures in implementing rating systems in the industry, the authors maintained that a system for evaluating and rating vendors is almost always essential. They finally recommended the use of the current capability of individual contractors for evaluation to the extent possible, instead of relying on past performance evaluation alone.

Barnaby and Bohannon conducted an investigation to determine the effectiveness of the pre-award survey (PAS) as an indicator of a contractor's ability to meet the delivery schedule (Barnaby and Bohannon, 1975). They recommended that information instruments such as the pre-award survey predictive index should not be instituted on an on-going basis, because such instruments would be used as evaluative indices. Also, if an individual were to be evaluated based on these pre-award survey predictive index numbers, the pressure would increase on the pre-award surveyor to favor an individual organization and introduce bias in the determination of the index.

Cormany and Donnellan developed some criteria for evaluation of contractor potential in the procurement of major weapon systems (Cormany and Donnellan, 1975). Schuman and Vitelli designed and performed a statistical experiment to evaluate certain indicators of contractor performance developed by the Air Force Logistics Command (Schuman and Vitelli, 1978). They concluded that deliveries appeared to be based on the capability of the contractor, and contractual requirements of delivery seemed to be irrelevant except that they were the best guesses available for contractor capability. Pingel proposed a system for evaluating service contractors (Pingel, 1981). McLennen outlined the feasibility of a decision support system for determining the criteria for source selection (McLennen, 1984).

From the literature it is clear that the move is towards best value procurement and away from the lowest cost bid criterion. However, there is no literature citing the use or development of domain-specific knowledge-based decision support for bidder evaluation in apparel manufacturing contracts, thus providing necessary justification for this project.

5.4 APPAREL MANUFACTURING AND QUALITY CONTROL

Literature in the area of apparel manufacturing and quality control was also reviewed to assess the effect of manufacturing technology and quality control practices on the overall capability of the enterprise and to serve as a means of acquiring knowledge for the AEEF.

5.4.1 Importance of technology on production and quality

Hodgins emphasizes the importance of higher levels of automation by stating that a non-automated process generally resulted in lower production and longer training periods for the personnel to achieve the desired level of product quality (Hodgins, 1990). Eberly describes the effect of technology on the apparel enterprise as follows (Eberly, 1990):

> New technology, in terms of both computer hardware and software as well as advanced spreading and cutting equipment, offers apparel manufacturers two significant opportunities to improve their companies' performances and response times.

The apparel manufacturing handbook by Solinger discusses basic production standards and information on all the operations involved in apparel manufacturing (Solinger, 1980). The stepwise details of the individual manufacturing operations and their required standards can be found in Hudson (Hudson, 1988). These sources strongly emphasize the importance of advanced technology on both quality and production rates.

5.4.2 Utility trouser manufacturing and quality control

The major official sources for the manufacturing and quality control of utility trousers are the military and federal specifications. These specifications have been developed for garments procured by DoD and other US Government departments. The military specification for utility trousers specifies the design, construction, stitches and seams, operations, tolerances and quality assurance provisions for the manufacture of utility trousers (MIL, 1984b). The military standard provisions for evaluating quality of trousers specifies standards for sampling, inspection and classification of defects (MIL, 1987). These specifications also refer to other military and federal specifications for buttons (V-B, 1984), fasteners (V-F, 1987), thread (V-T, 1982), label (DDD, 1987), cloth (MIL, 1984a), and sampling procedures (MIL, 1964). These specifications and standards can be utilized to evaluate the operations performed by the apparel manufacturing enterprise in producing utility trousers.

Based on the literature reviewed in the areas of source selection, apparel manufacturing and quality control and knowledge-based decision support systems, the following major conclusions can be drawn.

1. Only highly subjective vendor evaluation programs exist in DoD as well as the industry and there are no vendor rating programs. These evaluation programs are prone to the introduction of personal bias and consider past performance as the *only* major criteria.
2. There is no existing domain-specific knowledge-based apparel manufacturing enterprise evaluation system, with emphasis on new technology.
3. More recently, there has been a move away from the lowest cost bid approach towards a performance- or capability-based selection procedure.
4. Level of technology can be used as an important indicator of an enterprise's production and quality capabilities.

5.5 KNOWLEDGE ACQUISITION

The development of a **knowledge-based decision support system** for the evaluation of enterprise capabilities has been carried out in three stages: acquiring the knowledge, developing the knowledge framework and representing the framework as a computer-based system. The compiled knowledge existing in experts' minds and used in the evaluation process needs to be obtained in a structured format. Since the quality of knowledge framework heavily depends on the translation of the experts' evaluation skills into computer-representable knowledge, the knowledge acquisition process is critical to the development of the framework. As the first step in the knowledge acquisition process, the following three means were identified:

1. development and mailing of questionnaires to experts in the areas of apparel manufacturing and contracting, followed by analysis of responses;
2. knowledge from published literature in the fields of enterprise evaluation and apparel manufacturing technology and quality control;
3. interaction with experts.

5.5.1 Questionnaire to experts

The purpose of the questionnaire was to solicit experts' opinions on criteria that can serve as measures of a 'good' or 'ideal' manufacturing facility (Jayaraman, 1989b). With this objective in mind, eight major groups of factors were identified as being important for evaluating an apparel enterprise (Fig. 5.1). In addition, five major criteria for evaluating an enterprise's performance were identified (Fig. 5.2).

The questionnaire was divided into the following six parts:

1. rank performance criteria;
2. rank capabilities;
3. rank processes;
4. process descriptions;
5. experience with contracting;
6. company and personnel information.

The first three parts of the questionnaire were designed to obtain the relative importance and weights of various criteria. The fourth part, consisting of seven sections, dealt with specific questions about the following operational aspects of the enterprise:*

1. Raw materials and procurement practices;
2. Grading and marker making;
3. Spreading and cutting;
4. Sewing;
5. Packaging;
6. Shipping and distribution;
7. Quality assurance;
8. Miscellaneous items – information systems, organizational structure etc.

* Apparel manufacturing involves the following major process steps: grading of patterns and marker making; spreading and cutting of fabric; sewing of cut parts; inspection and packaging of finished garments; shipping of packaged goods.

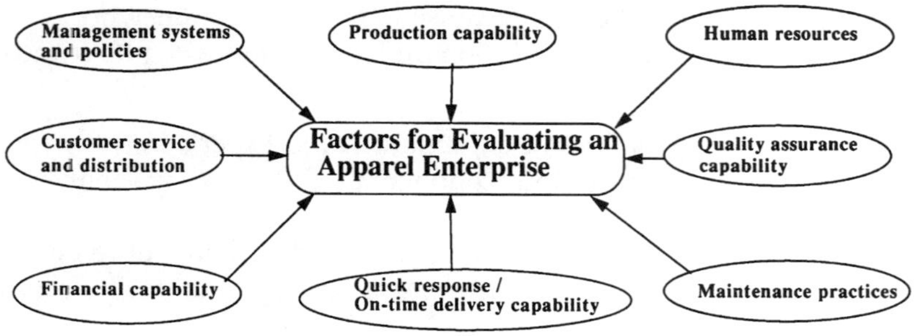

Fig. 5.1 Groups of factors for evaluating an apparel enterprise.

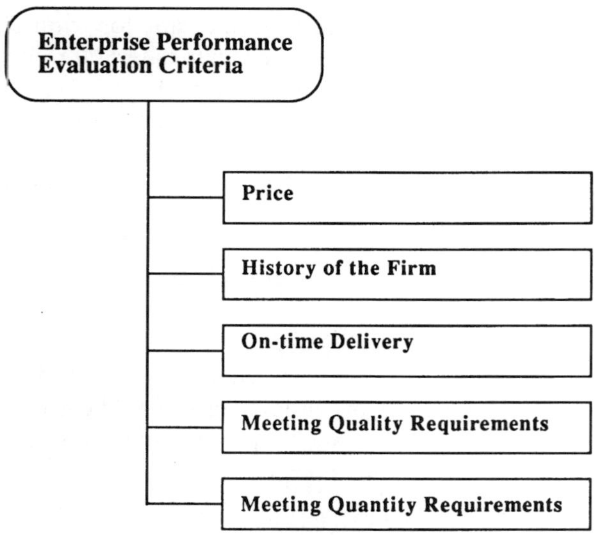

Fig. 5.2 Criteria for evaluating the performance of an enterprise.

These questions were expected to facilitate obtaining the complete list of factors and their relative importances for all the operations carried out, operator capabilities, machinery capabilities and so on, in a utility trouser manufacturing organization. The questionnaire was mailed to over 500 apparel companies through the American Apparel Manufacturers' Association (AAMA) and also sent to the Defense Personnel Support

Center (DPSC) and a few DLA field offices. The large number of questionnaires mailed out was to ensure that at least a few would respond; given the nature of the process, response from more than a handful was not expected.

The questionnaire asked the respondents to rank the importance of various factors determining the capability of the bidder, and also to rank the importance of the questions themselves. This ranking of questions was also critical in determining the weights for the various factors influencing the decision-making process.

5.5.2 Analysis of questionnaire responses

Eighteen responses were received for the questionnaires sent. The distribution of the response sources were as follows: three military agencies, four federal contractors and 11 general apparel companies. A statistical analysis was carried out to obtain the relative importance of the questions and weights for the various factors.* Based on the analysis, different factors for evaluating a bidder have been allotted points reflecting their relative importance. The questionnaire responses have been augmented with the help of available literature in the area of apparel manufacturing and quality control; the points were fine-tuned by discussing the results of the analysis with a panel of experts in the apparel industry.

The analysis of the questionnaire indicated a very high degree of importance for quality control and quality assurance, and the sewing and cutting operations. Among the various evaluation criteria, meeting quality requirements and on-time delivery were ranked first and second, respectively (Table 5.1). Quality assurance and production capabilities emerged as the factors having the maximum effect on performance (Table 5.2), and quality control and sewing were ranked as the most important processes (Table 5.3). The responses also clearly indicated that quality control activities should not be regulated by the production department. In other words, the quality control manager had to function independently of the production manager.

From the questionnaire responses, it is clear that quality control is the most important factor that is often evaluated for determining the capability of an apparel enterprise. Sewing, cutting and spreading are the three most important operations evaluated in the production of utility trousers. These operations are evaluated on the basis of several factors including the level of automation, machinery features, floorspace and operator capability.

* The findings were not subjected through a rigorous statistical validation process. Since our objective of the knowledge acquisition process was to elicit and collect 'subjective' data for implementation into a knowledge-based decision support system, and not just to draw conclusions based on statistical analysis, the number of responses was deemed adequate.

Table 5.1 Response summary for performance criteria (1 = highest rank, 5 = lowest rank)

Criteria	*Mean rank*	*Standard deviation*
Meet quality requirements	1.4	0.8
On-time delivery	2.0	0.7
Price	2.7	1.0
Meet quantity requirements	3.1	1.3
History of the firm	3.8	1.8

Table 5.2 Response summary for effect on performance (5 = maximum effect, 1 = no effect)

Criteria	*Mean effect*	*Standard deviation*
Quality assurance capability	4.8	0.5
Production capability	4.6	0.6
Human resources	3.9	1.0
Financial capability	3.5	0.9
Quick response capability	3.4	1.2
Management system	3.3	1.1
Customer service	3.1	1.2
Maintenance	2.9	0.8
Material handling	2.3	1.2
Warehousing and distribution	2.3	1.3

Table 5.3 Response summary for ranking of processes (1 = highest rank, 5 = lowest rank)

Process	*Mean rank*	*Standard deviation*
Quality control	1.7	1.0
Sewing	1.7	0.8
Cutting	2.3	1.2
Raw material inspection	3.0	1.9
Packaging	3.8	1.8
Shipping	3.8	2.4

5.5.3 Knowledge acquisition from literature

The effect of technology on the capability of the enterprise has been deduced primarily from the literature. The importance of floor area on the quality and efficiency of spreading and cutting is emphasized heavily by Jones (1990). He also provides information for determining the space requirements for the spreading/cutting room. The level of spreading machinery technology depends on the features of the spreading machines: automatic tensioning, end catcher etc. Solinger lists features that can be used for evaluating spreading machinery (Solinger, 1980). Quality control standards and tolerances for sampling and inspection are obtained from literature (MIL, 1964, 1984a, b, 1987).

For cutting machinery, different kinds of high technology systems such as numerically controlled (NC) knife cutting, laser beam cutting and water jet cutting are available. But no documentation is available indicating conclusively the superiority of one system over the other. One major conclusion that can be drawn from the literature is that a computer-controlled cutting system gives a higher production rate than manual cutting. In addition, a computer-controlled cutting system contributes to improved quality of the end product by producing more accurate cut parts. The sewing machines are also classified according to their contribution towards higher productivity and better quality, based on the level of technology.

The human resources in all the departments can be evaluated based on education, experience, wages, training, absenteeism, labor turnover etc. In the US, salaries of apparel workers vary considerably based on the geographic location of the plant. Hence, a scheme is required for normalizing the widely varying wages. The apparel plants wages survey divides apparel plants into seven groups by geographic regions (Apparel, 1989). It provides the average wages of all direct and indirect workers employed in an apparel enterprise in every region. The wage structures within each geographic region tend to be reasonably similar. Therefore, this survey has been used to normalize the wages. A constant multiplier is derived for each geographical region and the details of wages obtained from the apparel enterprise can be standardized with this multiplier.

5.5.4 Interaction with experts

In certain instances where it was not possible to obtain details either through the questionnaire responses or through the literature, discussions were held with experts in that specific area. For example, neither the questionnaire responses nor the literature yielded a reasonable range of annual labor turnover rate. The value was finally determined based on discussions with experts. Relative importances of most of the lower level criteria influencing the evaluation process were also determined with the help of experts.

5.5.5 Results of the knowledge acquisition process

Thus, the three-step knowledge acquisition process yielded the following results, which have then been used in building the knowledge framework:

1. indicators of enterprise performance on contracts;
2. abstraction of enterprise capabilities;
3. importance of procedures and processes in apparel manufacturing;
4. relative weights of various factors used in evaluation.

Quality, on-time delivery and price were identified as the three major indicators of contractor performance. Quality control and quality assurance practices emerged as the most important processes to be evaluated in determining the enterprise capability. Thus, the enterprise capability has been abstracted into quality, production and financial capabilities. These higher level abstract factors have been decomposed hierarchically to their subfactors until the specific subfactor becomes a parameter which can be observed or obtained from the enterprise being evaluated.

5.6 CHOICE SELECTION METHODOLOGIES AND SELECTION OF THE INFERENCE MECHANISM FOR THE DECISION SUPPORT SYSTEM

The design and development of this system falls into the category of choice selection procedures for ranking various alternatives based on different criteria with different weights. Multidimensional scaling and multi-attribute decision-making are two of the major techniques available for selection or rating of alternatives. Literature in these two areas has been reviewed and considered for implementation. Also, implementing this knowledge as an uncertainty management system for evaluation of manufacturing enterprises, based on the Dempster–Shafer theory of probability has been considered (Rothman, 1989; Shafer, 1976). This implementation would have resulted in a a probabilistic ranking of several alternatives. A fuzzy set formulation of this decision support system has also been considered (Zadeh, 1965).

The inference mechanism is crucial for the system to manipulate the knowledge base and arrive at the results. The information pertaining to the enterprise being evaluated is stored in the knowledge base and it must be manipulated by the inference mechanism to compute the rank or index of its capability to perform on a contract. Here the inference mechanism can be viewed as a method of ranking the specific enterprise in a pool of competitors and hence is also referred to as the **evaluation function**. Four techniques have been considered in developing the evaluation function for ranking the capability of the apparel enterprise. They are:

1. multidimensional scaling;

2. multi-attribute decision-making techniques;
3. probability techniques;
4. polynomial function.

The following sections examine the feasibility of applying these techniques to AEEF.

5.6.1 Multidimensional scaling

The multidimensional scaling technique uses a scaled rank for each attribute and the attribute's contribution towards the criteria being evaluated, to arrive at the multidimensional representation of the various candidates. The result is the positioning of each of the candidates in a space of n dimensions, where n is the number of factors/attributes considered. However, an evaluation function is needed for computing the overall rating. The details of this technique can be found in several papers and textbooks (Green and Carmone, 1972; Green and Rao, 1972; Green and Wind, 1973). Though no work has been reported on applying multidimensional scaling to vendor evaluation or similar problems, Green and Wind (1973) have demonstrated the feasibility of developing such an evaluation function. Ideally this would be the best evaluation strategy for a one-time evaluation process. But, for the AEEF it is inappropriate for the following reason:

1. The result is the positioning of the various candidate enterprises in scales of different attributes. The overall preference rating still needs to be calculated by some evaluation function.
2. The matrices to be manipulated are of dimensions $n \times n$, where n is the number of factors. In the AEEF the number of factors will total hundreds.
3. There is no possibility of a hierarchical grouping of various factors.

Also, the reliability of this evaluation function is assured only if a large number of evaluators/respondents is used to assign the preference rating. Hence this technique is mostly suitable as a knowledge acquisition methodology for determining the weights of various attributes in an evaluation function. However, this technique has not been used in determining *weights* for the AEEF, since a simple statistical analysis could serve the purpose.

5.6.2 Multi-attribute decision-making techniques

One of the multi-attribute decision-making methods for obtaining the relative ratings of more than two candidates, is the eigenvector method (Liu *et al.* 1990). In the eigenvector method, one or more evaluators assign the preference rating between pairs of the candidates on various attributes to form the matrix of preference rating entries. Each evaluator will have one preference rating matrix for each attribute. The eigenvectors of each of these matrices give the relative rating of each of the candidates. Hwang and Yoon 1981) provide complete details of the eigenvector method.

From the AEEF standpoint, the eigenvector method has the following disadvantages.

- The number of pairs needed for rating n candidates is n_{C_2}, which becomes very large even for a reasonable number of candidates.
- The statistical reliability (and hence quality) of the result is related to the number of evaluators; therefore a large number will be required.
- It does not take into account the relative importance (weights) of the factors contributing to the decision-making parameter.

Though the number of pairs needed for evaluation can be reduced by a statistical sampling technique developed by Smith (BestChoice 3, 1990), the method becomes impractical for the AEEF, owing to the other two reasons previously mentioned.

5.6.3 Probability techniques

The bidder score can also be probabilistically determined from the bidder data. This scheme will provide the bidder score with confidence levels. For example, bidder A is assigned a score X with confidence level Y. The confidence level is a measure of the uncertainty introduced in the inference process due to unavailability of some required data. An average value can be substituted in the evaluation function for the missing data, but the confidence level will be reduced by a fraction. The reduction in confidence level will be proportional to the importance of the missing data. An important thing to note in this method is that another evaluation function is still needed for ranking the bidder.

There are some conceptual problems with the average value substitution and reduction of confidence level. As the system confronts unknown, critical data the confidence level becomes zero or near zero. A further reduction of confidence level is then impossible. Also, it is not possible to have a decision rule combining the ranking/score and the confidence level. This might lead to additional problems in decision making rather than simplifying it. For example, a bidder who deserves a 10% score can obtain a score greater than 50% by not providing certain data. However, obtaining a higher score in this way is possible only at the cost of reduced confidence level.

5.6.4 Polynomial function

A polynomial function can be used for evaluating any higher level entity based on the rank/score of the contributing lower level entities (Fig. 5.3):

$$Y = \sum_{i=1}^{n} X_i \tag{5.1}$$

where Y is the Score of the higher level entity; n is the number of subclasses contributing to the higher level entity; x_i is a polynomial function (of degree m) of the i_{th} entity's Score (x_i), which is of the form

$$x_i = \sum_{j=0}^{m} w_{ij} \cdot x_i^j \tag{5.2}$$

and w_{ij} are the Weights of the j_{th} power term in the polynomial for the i_{th} entity.

The polynomial ranking function is complete as well as complicated. However, it will be extremely difficult to get the values for all the coefficients w_{ij} for higher powers of x ($j \geq 2$) either from the questionnaire responses or from discussions with experts. Hence, for this system the polynomial has been simplified to a simple linear evaluation function of the form

$$Y = \sum_{i=1}^{n} w_i \cdot x_i \tag{5.3}$$

where

Y is the *Score* of the higher level entity, n is the number of lower level entities contributing to the higher level entity, x_i is the *Score* of the i_{th} entity and w_i is the *Weight* of the i_{th} entity.

5.6.5 Working of the inference mechanism

The linear evaluation function given by equation (5.3) has been chosen for the inference mechanism. In the actual implementation, however, some offspring classes contribute to more than one parent classes. This multiple inheritance, therefore, leads to a minor modification in the evaluation function, which is the addition of the parent index. The Score (Yk) of the higher level class (k) to be evaluated is on the left-hand side, where k is an index for the offspring to indicate that the class is the k_{th} parent. When there is no multiple inheritance, i.e., when all offspring has no more than one parent, the significance of the index k is ignored.

$$Y_k = \sum_{i=1}^{n} w_{ki} \cdot x_{ki} \tag{5.4}$$

The variables in the function are the *Scores* of the offspring nodes (x_{ki}) and the coefficients are their respective *Weights* (W_{ki}) towards the object/class under consideration (k).

5.7 DESIGN OF THE KNOWLEDGE FRAMEWORK

The knowledge obtained from experts and literature for the development of the AEEF is hierarchical in nature. Therefore, an object-oriented representation technique is well suited to represent the knowledge in a knowledge-based system. The factors used as criteria for evaluating the apparel enterprise are represented as **classes**. A hierarchical graph structure is followed for the successive decompositions of the classes into their subclasses (Fig. 5.3). With a few exceptions,* all the properties of the parent class are inherited by the offspring; however, the values of the properties are not inherited. An instance of a class (factor) is represented as an **object**. Every class and object belonging to a class has two basic properties: *Weight* and *Score*. Figure 5.3 shows the two basic properties inherited from the level 1 class *Overall_Score* by all its subclasses.

The property *Weight* is a decimal fraction value represented as the relative importance of that class with respect to *Weights* of all its sibling classes. Hence the sum of *Weights* of all the offspring of any class must always be 1. The property *Score* represents a ranking value calculated for that class from its subclasses. The *Score* varies between 0 and 4, with 0 being the lowest and 4 the highest Score. During the start of an evaluation session, the *Score* of all classes is set to 0. The *Score* for the lowest level class is calculated on the basis of appropriate heuristics which act on the other properties or features of that lowest level class. For example, if the sewing machine for producing pockets has an automatic positioning feature it will get the highest *Score*; on the other hand, if it only has cam control, it will get the next lower *Score*, and so on. This *Score* is used in determining the *Score* of the next higher level class and propagated upwards. This upward propagation of the *Score* will continue until the highest level, i.e. the *Overall_Score* of the bidder is determined.

Though the major portion of the class hierarchy graph is a tree, in certain classes, multiple inheritance occurs from more than one parent. For example, manufacturing features is a factor that will contribute to both the production capability and the quality capability of a facility. Consequently, there will be more than one *Weight* and *Score* associated with that child class, and these *Weights* and *Scores* will be indexed in order, for correct propagation to the right parent. For example, if the class

* These exceptions are only implementation dependent and not conceptual. When the children do not use all the inherited properties, the properties not used are manually pruned and only the needed properties are retained. This is to compensate for the nature of the implementation vehicle, Nexpert Object, which creates copies of the parent properties in the offspring, instead of dynamically inheriting them.

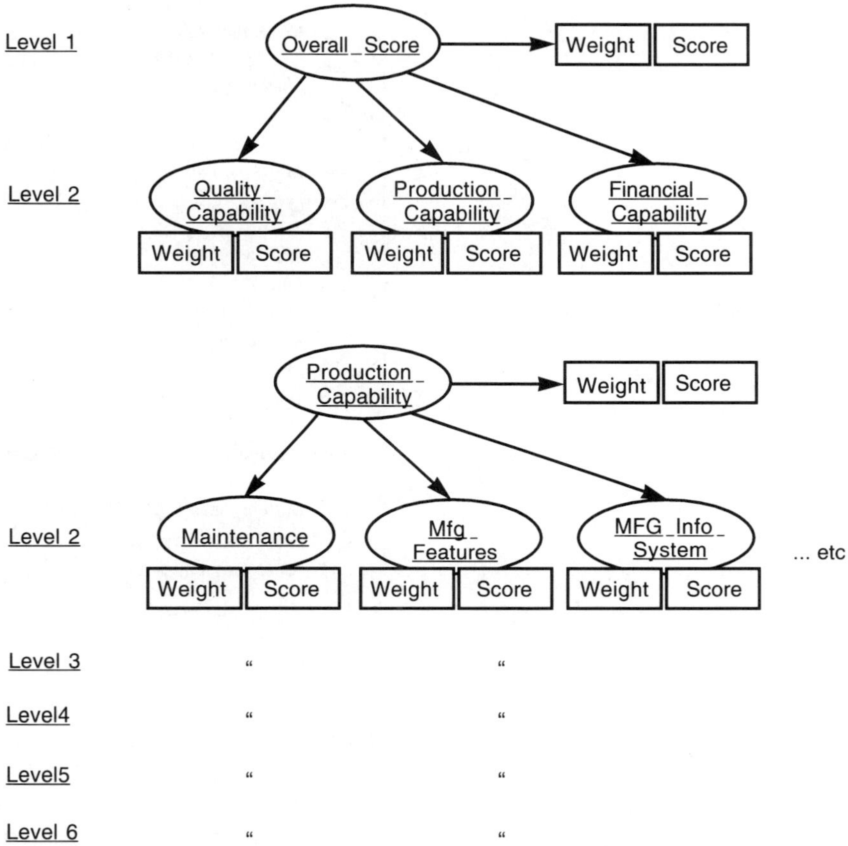

Fig. 5.3 Inheritance of properties by lower level classes.

Mfg_Features contributes towards both *Quality_Capability* and *Production_Capability*, it will have the properties *Weight* and *Score* for its propagation towards the class *Quality_Capability*, and *Weight1* and *Score1* for its propagation towards the class *Production_Capability*.

5.8 THE KNOWLEDGE FRAMEWORK

As mentioned in section 5.5, *Quality_Capability*, *Production_Capability* and *Financial_Capability* were identified as the three main factors for evaluating a bidder's facility. Thus, these three classes – at level 2 – contribute to determining the *Overall_Score* for the bidder at level 1, the highest level (shown in Fig. 5.4).

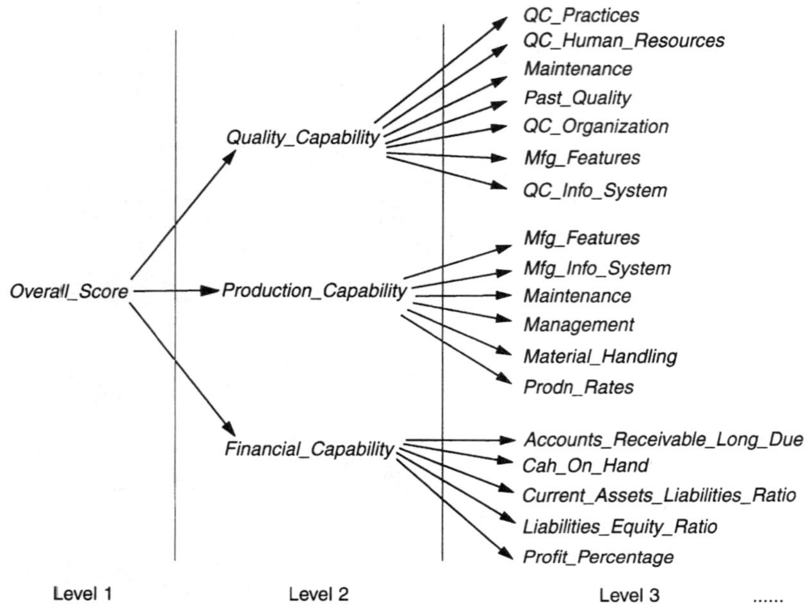

Fig. 5.4 Decomposition of the class *Overall_Score.*

The sets of factors, according to which the questionnaire was framed, are grouped under these three level 2 classes. For example, the factors human resources and maintenance can be part of *Quality_Capability* as well as *Production_Capability*, as they contribute to both. Similarly on-time delivery is a result of good quality and production capabilities. So, these factors are subsumed by both the classes *Quality_Capability* and *Production_Capability*. Distribution and management policies received very low relative importance in the questionnaire responses; therefore, a separate class has not been created and the factors have been included as part of *Production_Capability*.

The three level 2 classes – *Quality_Capability*, *Production_Capability* and *Financial_Capability* – have been further decomposed hierarchically to identify the important subfactors that contribute to determining their values. The next level classification of the factors under *Quality_Capability*, *Production_Capability* and *Financial_Capability* is also shown in Fig. 5.4.

Quality_Capability was considered to be the most important factor by 83% of the respondents to the questionnaire. *Production_Capability* was the next important factor. The relative weights of these three classes have

Table 5.4 Distribution of Weights

Overall_Score	
Quality_Capability. Weight	0.45
Production_Capability. Weight	0.35
Financial_Capability. Weight	0.20
Quality_Capability	
QC_Practices. Weight	0.27
QC_Human_Resources. Weight	0.18
QC_Info_System. Weight	0.10
QC_Organization. Weight	0.10
Mfg_Features. Weight	0.27
Maintenance. Weight	0.06
Past_Quality. Weight	0.02
Production_Capability	
Mfg_Features. Weight1	0.50
Prodn_rates. Weight	0.15
Mfg_info_system. Weight	0.10
Maintenance. Weight1	0.10
Material_Handling. Weight	0.06
Management. Weight	0.09
Financial_Capability	
Accounts_Receivable_Long_Due. Weight	0.10
Cash_On_Hand. Weight	0.40
Current_Assets_Liabilities_Ratio. Weight	0.30
Liabilities_Equity_Ratio. Weight	0.10
Profit_Percentage. Weight	0.10

been arrived at by proportionately distributing the relative importances of the factors grouped under the three classes. Table 5.4 shows the weights of all classes (up to level 3), eventually contributing to the *Overall_Score*. (These weights can be easily modified by the evaluators to suit the requirements on a specific procurement.) In this chapter, only the classes up to level 3 are explained. A complete listing of the weights and a full decomposition of the knowledge framework can be found in Narayanan (1991) and Narayanan, Olson and Jayaraman (1994).

5.8.1 Decomposition of the level 2 class *Quality_Capability*

Figure 5.4 shows the various subfactors under *Quality_Capability*. Of these, *QC_Practices* has been identified as one of the important factors determining the *Quality_Capability* of an apparel enterprise. *QC_Practices* encompasses the quality control checks to be performed at every stage of the process, from raw material inspection to packaging. The evaluation process for determining the efficiency of the QC checks performed is complicated for at least three reasons.

1. The number of QC checks that can be instituted at every stage of the manufacturing operation is fairly large.
2. Certain QC checks need not be performed in a specific enterprise either due to process differences or complete elimination of those corresponding defects.
3. Getting complete information from the bidder about all QC checks is very difficult from both the bidders' side and the evaluator's side.

Therefore, a scheme has been devised in which a select number of important QC checks is listed and the bidder can specify whether those QC checks are being performed in the facility. The list of selected QC checks has been compiled on the basis of discussions with experts and results of a survey on apparel defects analysis (Srinivasan, Dastoor and Jayaranan, 1992). A minimum number of QC checks should be performed to obtain a score equal to the minimum score. On the other hand, not all QC checks need to be performed to obtain the maximum score, i.e., a major subset of the QC checks would be sufficient to obtain the maximum score. Therefore, any facility performing more than the built-in threshold upper limit of the number of QC checks to be performed will be assigned the maximum score. The need to perform raw material inspection checks is also waived if the bidder buys the raw material (except fabric, for which there is no acceptable suppliers list) from a supplier approved by DoD.

Manufacturing features

The class *Mfg_Features* is one of the most important of the level 3 classes. It covers both the features of the production machinery, and the production personnel. Good machinery and an experienced and efficient work force largely determine the production and quality capabilities of the facility. Hence, *Mfg_Features* contributes to both *Production_Capability* and *Quality_Capability* as shown in Fig. 5.5. Moreover, it has been considered to be an equally important factor as *QC_Practices* in contributing to *Quality_Capability*.

The maintenance of quality in the QC department depends heavily on the QC personnel, and therefore *QC_Human_Resources* has been chosen as the third most important factor under *Quality_Capability*. Fig. 5.5 shows the categories and attributes of personnel under the QC department that will be considered in the evaluation process.

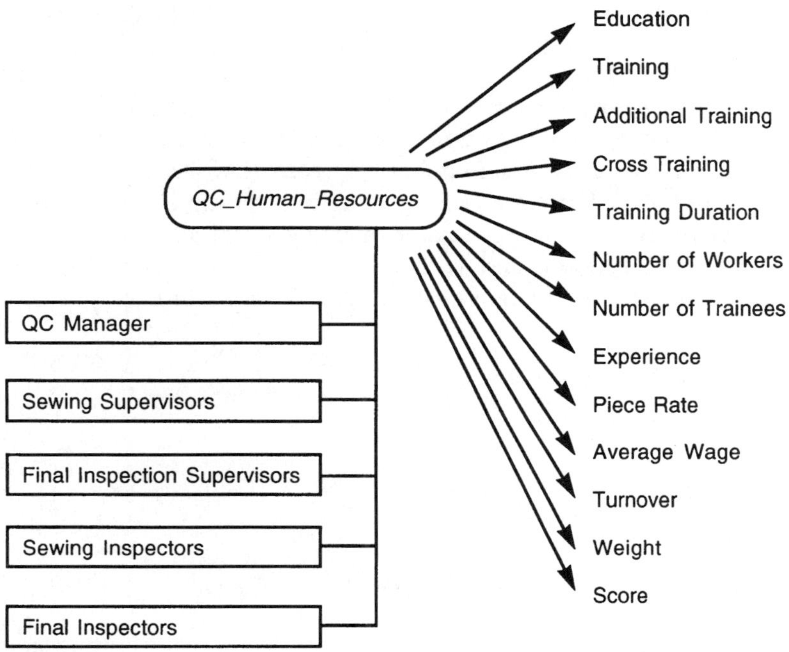

Fig. 5.5 Categories and attributes of personnel in the QC department.

Past_quality evaluation

In normal practice, past quality performance of the bidder is the primary factor in the evaluation procedure. However, in the informed knowledge-based approach, where the analysis of the data obtained from the bidder's facility gives a more reliable and accurate estimate of the capabilities, the past quality performance can be regarded as one of several factors. This is indicated by the low weight (2% of 45%) assigned to the class *Past_Quality*. Nevertheless, the past quality score needs to be calculated from the estimate of the evaluator. A simple procedure has been developed that takes into account the evaluator's estimate of the past quality performance of the bidder and the number of years the bidder has been in business.

During the evaluation of past quality performance, a score can be assigned by the evaluator for the number of years the bidder has been in business. If the bidder has been in business for a sufficiently long time (represented by the class *OK_Years_in_Business*), then this score itself would be appropriate as the *Past_Quality* score. On the other hand, if the bidder has been in business for only a short time, then the evaluator can assign a score considering only the known time frame, i.e. the number of years the bidder has been in business (represented by the class *Years_in_Business*). A weighted average of the assigned score and an average score (2.0) is assigned for the unknown number of years (*OK_Years_in_Business* – *Years_in_Business*). Since a higher confidence level can be associated with the score assigned by the evaluator over the weighted average score, a weight of 80% is associated with the assigned score and 20% for the unknown years' score to arrive at the *Past_Quality* score:

$$s_{\text{PastQuality}} = s_{\text{Eval}} \times 0.8 + \frac{s_{\text{Eval}} \times Y_{\text{Known}} + 2.0 \times Y_{\text{OK-Known}}}{Y_{\text{OK}}} \times 0.2 \tag{5.5}$$

where S_i are the *Scores* and Y_i are the Years.

5.8.2 Decomposition of the level 2 class *Production_Capability*

Mfg_Features is the single most important factor contributing to the *Production_Capability* of an apparel enterprise (Fig. 5.4, Table 5.4). Next, the production rates (represented by the class *Prodn_Rates*) are important in determining the *Production_Capability*. An efficient production department also needs effective and well-designed information systems (*Mfg_Info_System*) and material handling systems (*MH_System*). Complementing these features are the *Maintenance* and *Management* as subclasses of *Production_Capability* in Fig. 5.4. The effectiveness of management policies is determined on the basis of whether there were strikes or lockouts in the past, any bonus was given to the employees in the past and the enterprise were unionized.

Manufacturing features

Mfg_Features has been divided into *Mfg_Human_Resources* and *Machinery_Features*. The class *Mfg_Human_Resources* is very similar to *QC_Human_Resources*. Apart from the standard attributes for human resources, three additional attributes (the number of utility operators, the number of operators who can sew the seatseam and the number of operators who can sew any felled seam) have been defined for evaluating manufacturing human resources. (Seatseam and felled seam are the two critical operations in utility trouser manufacturing and have been selected as representative operations for evaluating the capability of manufacturing human resources.)

Grading and marker-making machines, spreading machines, cutting machines and sewing machines are the four types of machinery considered in the class *Machinery_Features*. The number of machines in each category along with their capabilities determine the score of these classes. Computerized grading and marker making, along with numerically controlled cutting machines, help in achieving higher productivity and quality; therefore they are given maximum scores in their respective categories.

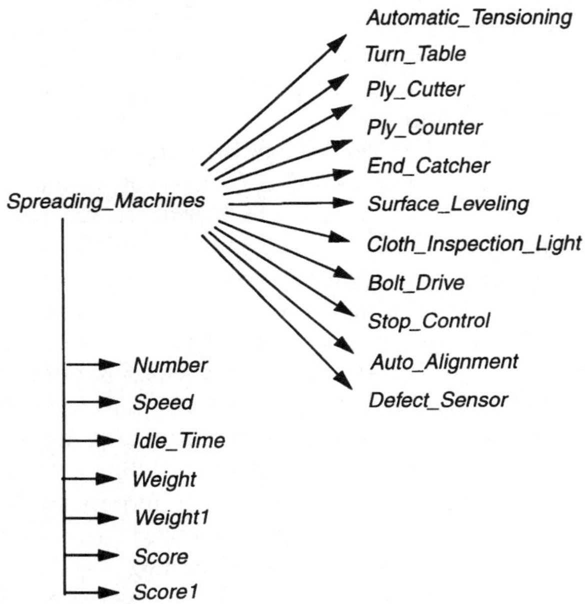

Fig. 5.6 Spreading machine features and their attributes.

Spreading machines are evaluated based on the features they possess (Fig. 5.6). For example, the existence of an automatic tensioning device will enhance the *Score* of the class *Spreading_Mach*, and if the spreading machine possesses all the features listed, it will result in the maximum *Score* of 4.

Sewing machine classification

For most manufacturing operations, modern machinery incorporating higher levels of technology tends to reduce the proportion of defective units produced in a factory and causes fewer quality problems at higher production rates. But there is also a tradeoff for the increase in technology level, as it is costly, and the traditional measures of returns on investment (on a higher level of technology) start to diminish after a certain point. So it is crucial to identify, for every operation, the best technology level and the technology level that would be the minimum requirement.

In an apparel manufacturing facility, sewing is the most important operation, and adds the maximum value to the fabric in its transformation into a garment. It would be sufficient if we could develop a good scale for the technology level required for a sequence of operations to produce the garment, in this case, the utility trouser. But this process is quite difficult since the change in technology level often necessitates combining or splitting certain operations, and altering the sequence in which they are carried out. So it is necessary to have different sequences of operations, which represent the range from the best technology level to the worst.

To develop these sequences of unit operations, an important prerequisite is a scheme for the classification of technology levels. Sewing machines need to be classified into a specific order of technology levels based on their features and capabilities. A parameter *Feature_Number* is defined to represent the technology level (the more advanced the technology, the higher the *Feature_Number*). The features and the classification of sewing machinery based on these features are given in Table 5.5. Also, a database of various sewing machinery available in the market has been developed. It contains the manufacturer name, model name and number, a brief description of the machine, the technology level of the machine with supporting reason for the classification, the stitch type, cost, operating and maximum speeds, space occupied and the training time required for operators and mechanics. This database was useful in estimating the space occupied by sewing workstations. It will also be useful in evaluating the sewing machinery available in the facility, and it should be expanded and updated frequently in order to reflect the frequent changes in technology.

Three sequences of operations representing the worst-case, mid-case and best-case technology levels have been developed for the production of utility trousers (Jayaraman, 1989b). The best-case technology level for a certain operation is the one beyond which a higher technology level is either not essential or does not contribute to a higher quality or faster rate of production. Hence, a technology level higher than the best-case values are treated as being the same as the best-case values for that specific operation. These best- and worst-case technology level sequence values have been used in the evaluation process as the optimum and worst *Feature_Numbers*, respectively. Also, the best-case technology level sequence of operations has been used for determining the relative weights of some selected sewing operations (discussed in the following section).

Table 5.5 Technology level classification of sewing machinery

Feature[a]	*Feature*	
1	Basic machine	
2	Threadtrimmer	OR
	Undertrimmer	OR
	Felling folder	
3	Cam control	OR
	Electronic motor control	
4	Automatic workaids	
	e.g. automatic belt loop cut and count	OR
	automatic feed	OR
	button sew	
	OR	
	Programmable electronic motor	
5	Multifunction programmable	OR
	Fully automatic	

[a] The better the features, the higher the feature number.

Weight determination for sewing operations

There are 23 unit operations in producing the utility trouser (MIL, 1984b); these are shown in Fig. 5.7. Since it may be difficult and even unnecessary to obtain information on machines used in all these operations, ten important operations have been identified. The importance of an operation is measured by the relative weight assigned for that operation. The criteria for determining the relative weights for the operations are as follows.

1. The operation must be one of the most critical.
2. The technology level of the machine required for that operation must be very high.

However, currently available technology levels of sewing machines for some of the critical operations are low. Hence, a combination of the criticality of an operation and the highest technology level of the machinery possible for that operation has been used as the relative weighting factor for each of the ten operations. Proportionate weights are given to the technology level component as well as the criticality of operation component, which between themselves are weighted in the ratio 40:60. These component weights are added together to get the weight of the individual operations. The selected ten operations, each

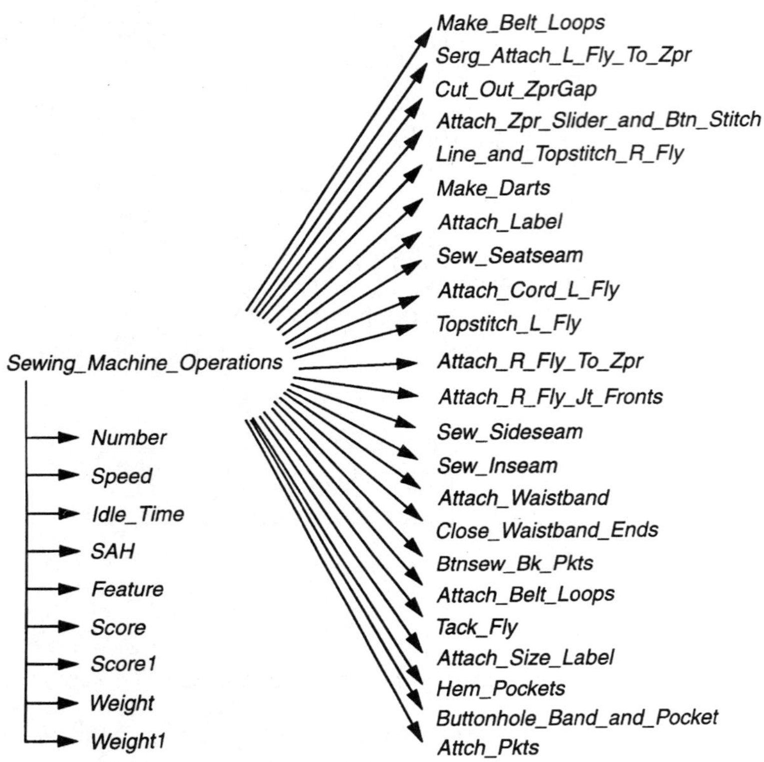

Fig. 5.7 Sewing machine operations and their attributes.

with its respective criticality rank, best technology level possible (*feature number*), weight for technology level and criticality, and the final *weight* calculated, are shown in Table 5.6.

Maintenance

The effectiveness of maintenance is evaluated by the procedures and human resources used in maintenance. Also, the availability, and reliability of the machines can be a good indicator of the condition of the machinery.

The *Reliability* of the machines can be evaluated by the two standard parameters:

- mean time between failures (*MTBF*);
- mean time to repair (*MTTR*).

Table 5.6 Determination of weights for sewing operations

Operation	*Best possible technology level feature no.*	*Technology level weight*	*Criticality rank*	*Criticality weight*	*Weight*
Attach belt loops	4	0.04	4	0.04	0.08
Attach label	5	0.05	5	0.02	0.07
Attach pockets	4	0.04	4	0.04	0.08
Attach waistband	4	0.04	3	0.06	0.10
Make belt loops	4	0.04	4	0.04	0.08
Make darts	5	0.05	3	0.06	0.11
Sew inseam	3	0.03	2	0.09	0.12
Sew seat-seam	3	0.03	1	0.12	0.15
Sew side-seam	3	0.03	2	0.09	0.12
Topstitch back darts	5	0.05	4	0.04	0.09
Total		0.40		0.60	1.00

These two parameters can be obtained individually for the spreading, cutting and sewing machines. Also, a higher number of *Spare_Sewing_Machines* can improve the efficiency of the sewing department and thereby contribute to increased *Reliability*.

Material handling

Automated material handling systems, such as automated guided vehicles (AGV) and unit production systems (UPS), can reduce the material idle time and thereby improve the production capability of an apparel enterprise. Also, the ease of material handling is determined by the amount of space available per machine. A very high machine area to total area ratio indicates insufficient material handling space, whereas a very low ratio indicates a lot of wasted space.

Production rates

Sewing is the most labor intensive and most important of the various steps in apparel manufacturing. Moreover, it tends to be the principal factor affecting the output of the enterprise. Consequently, assessment of the sewing capacity will provide a good indication of the bidder's production rates and hence the capacity. Therefore, only *Sewing_Capacity* is taken into account while evaluating *Prodn_Rates*. The decomposition of *Sewing_Capacity* is very similar to the decomposition of *Sewing_Mach*, where only the selected ten sewing operations have been used in the evaluation. The sewing capacity of each of these operations is estimated on the basis of the comparison of standard allowable hours (*SAH*) for that operation, number of sewing machines allocated for that operation, working hours per day and the number of trousers to be assembled per day.

Manufacturing information systems

The bidder's inventory control system, scheduling system, maintenance of production records and traceability of order status to the departments and to the individual sewing machines have been chosen as the four most important factors determining the effectiveness of the manufacturing information system. The existence of separate control systems for fabric, trim (buttons, zippers, thread etc.), finished goods and other supplies determines the effectiveness of the *inventory control system*. The purchase lead times for the various raw materials are indicators of the effectiveness of the inventory control system. The *Scheduling_System* is evaluated on the basis of the schedule update frequency and computerization of the process. The existence of cut order delivery performance records, and their being on-line are also used as important criteria for evaluation of the manufacturing information systems.

5.8.3 Decomposition of the level 2 class *Financial_Capability*

The class *Cash_On_Hand* is a crucial indicator of the financial status of the company (Fig. 5.4). Too little or too much cash indicates an unhealthy situation. If a major portion of the accounts receivable is long overdue (more than six months), the chance of collecting it becomes remote and hence is viewed negatively. Also, values of current assets vs. liabilities ratio, liabilities vs. equity ratio and profit percentage should neither be too low nor too high. Thus, these five factors have been considered in evaluating the financial capability of the enterprise (see Fig. 5.4 for details).

5.9 SOFTWARE IMPLEMENTATION OF DECISION SUPPORT TOOL

Automation is one of the important keys to productivity. The development of the knowledge framework for the evaluation of apparel enterprises (the AEEF) leads to the next logical step of automating the evaluation process. However, total automation of the evaluation process may not be the best solution. There are numerous unknown factors which, in addition to the capabilities evaluated by the framework, may influence the decision-making process. Hence the framework should be used to automate the evaluation process only to the extent that it serves as a decision support tool for the human evaluator. Therefore, the knowledge-based framework has been implemented as a decision support system, with an appropriate implementation vehicle.

5.9.1 Selection of software implementation vehicle

A hierarchical object-oriented representation technique has been adopted to represent the knowledge acquired in a computerized system (Narayanan, Olson and Jayaraman, 1994). The AEEF also consists of a large number of rules which act on the information about the apparel enterprise to determine its capability. Hence, a hybrid of object-oriented representation and rule-based inference strategy is required for an efficient implementation of AEEF as a knowledge-based system. The object-oriented expert system shell 'Nexpert Object' has been selected for the implementation, since it supports both object-oriented representation of knowledge and rule-based reasoning strategies to act on the objects. Another requirement is that the resulting knowledge-based system be available on MS-DOS, as well as the UNIX operating system, so that it can be used by a large number of people. Consequently, the availability of Nexpert Object on both UNIX and DOS operating system environments has been a major factor in its selection as the implementation vehicle. Srinivasan performed a comparative study of Nexpert Object, other expert system shells and traditional programming languages (Srinivasan, 1991). He discussed the advantages of Nexpert Object as a knowledge-based system development tool in terms of faster prototyping, easy linkage to other languages and databases, availability on various platforms etc. He recommended Nexpert Object for an efficient implementation of object-oriented and rule-based hybrid knowledge-based systems. He also made an economic justification for the selection of Nexpert Object, when compared with other expert system shells.

5.9.2 Implementation of the knowledge framework

The development version of Nexpert Object provides a graphical representation tool which contains a set of form-based editors (Nexpert

Object, 1988a,b). Different editors are available for creating and editing classes, objects, properties and rules. When these form-based graphical editors are filled, the system automatically generates the code in ASCII format, which is portable across UNIX, MS-DOS and Macintosh platforms. The knowledge base and the inference engine for bid evaluation software tool (BEST) have been created with the help of Nexpert Object's form-based editors.

The conceptual framework of criteria for evaluation has been represented as a hierarchy of Nexpert Object **classes** (Narayanan, Olson and Jayaraman 1994). The Nexpert Object class can have **subclasses** as well as **properties**. The individual contractor details are represented as **objects**, which are created as instances of the classes defined. The attributes of the contractor details are represented as **properties** of the classes.

The knowledge base is mainly composed of a set of **if-then** type production rules. The rule has a **condition** part which is verified by the '*if*' clause, and a **hypothesis** part which is set to **true** if the condition is satisfied and **false** if the condition is not satisfied. The rules are identified by a unique rule number and are alphabetically ordered according to the hypotheses. The rules also have an **action** part on the right hand side, which triggers additional knowledge processing or data alterations, if and only if the hypothesis becomes *true*. A sample rule is given in Fig. 5.8. If a rule has to be fired, data required by the condition part of the rule should be provided to the system. The process of supplying data required by a rule for its firing is known as **volunteering**.

There are two types of rules in the AEEF: **knowledge rules** and **propagation rules**. The lower level rules are the *knowledge* rules, which compute the *Score* of the lower level objects from the properties and values of the lower level enterprise details. For example, if the number of knots or splices in 1000 m of the sewing thread is less than or equal to one, then a *Score* of 4 is assigned to the class *Knots_Splices*. The first rule in Fig. 5.8 (R11) is also an example of a knowledge rule. The higher level rules are the *propagation* rules, which propagate the *Score* from lower level objects to a higher level object. These propagation rules are part of the inference mechanism, but they do not compute the scores from the AEEF's knowledge network. They just derive the score of higher level objects from the *Scores* and *Weights* of the lower level objects. The second rule (R481) is an example of a propagation rule.

The calculation of the *Score* of a higher level object with the propagation type of rules requires that the *Score* of the lower level objects be already computed. These precedence constraints impose a sequence for the firing of the rules. This sequence is established by modifying the properties of the **meta-slot** of the rule hypothesis (Figure 5.9). The meta-slots have many properties such as the inference category number, initial values, inheritance strategies and prompt line, which can control the inference

```
(@RULE-R11
  (@LHS=
    (>(|Attch.Waistband|.Feature-WorstAttch.Waistband) (0))
    (>(|Attch.Waistband|.Feature-BestAttch.Waistband) (0))
  )
  (@HYPO=Atch.W.Band)
  (@RHS=
    (Do (4*(|Attch.Waistband|.Feature-WorstAttch.Waistband)/
          (BestAttch.Waistband-WorstAttch.Waistband))
       (|Attch.Waistband|.Score))
  )
)


(@RULE=R481
  (@LHS=
    (>=(LotSize)        (0))
  )
  (@HYPO=ThreadStdsScoreDtmd)
  (@RHS=
    (Do
      (|ThreadElongation|.Weight *|ThreadElongation|.Score +
        |KnotsSplices|.Weight * |KnotsSplices|.Score)
    (|ThreadStds|.Score))
  )
)
```

Fig. 5.8 Sample knowledge and propagation rules.

process. For instance, when the inference category number is used to control the order of firing of the rules, the lower the value of the number, the later the rule will be fired. Thus, if rule A requires the result of rule B, rule B will have a greater inference category number than rule A, and consequently, rule B will be fired first.

Another important use of meta-slots is for the initialization of the *Scores* without having additional rules. This is achieved by setting the *InitValue* of the meta-slot of the highest level class *Overall_Score*'s property, *Score*, to 0 and propagating the values to all the subclasses. The meta-slots are also used to control the inheritance strategies. For example, the inheritance of *Score* to all lower level classes should take place only for the initial value of zero. Subsequently, when knowledge processing takes place, the values should not propagate downwards. Otherwise, all the lower level scores would be lost. The meta-slot properties stop the inheritance of values once knowledge processing

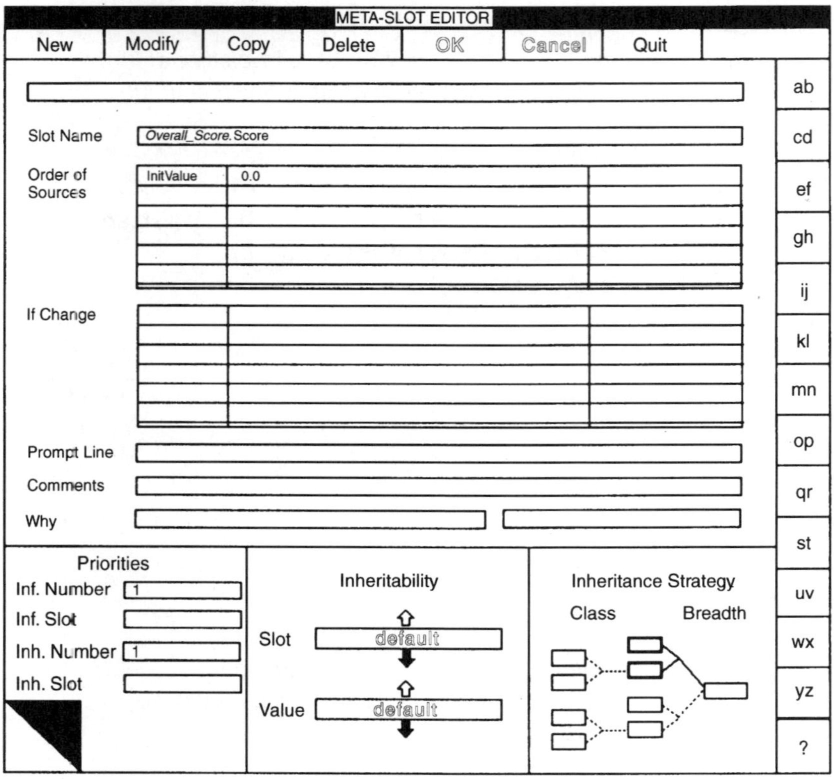

Fig. 5.9 Inference strategy control through meta-slot editor.

starts. The meta-slots also control user interaction with the system during knowledge processing. The system's prompts asking the user to input the values of objects, can be modified by altering the *Prompt Line* field. For example, the system can ask the user to 'Enter the Master Data File Name:' instead of asking 'What is the value of *File_Name*?', by modifying the meta-slot *Prompt Line*.

5.9.3 User-modifiable decision variables

The knowledge processing mechanism makes use of various decision variables derived from the knowledge framework. These decision variables are subjective in nature. Hence for different needs, these decision variables may need to be modified. Also, the present standards or specifications may become obsolete. For example, when new specifications that supersede the current military specifications for manufacturing and quality control of utility trousers are issued, some

```
Max_Fabric_LT                 8      Weeks
Max_Trim_LT                   4      Weeks
Max_Spare_Parts_LT            6      Weeks

Min_Cutting_MTBF              100    Hours
Min_Sewing_MTBF               50     Hours
Min_Spreading_MTBF            100    Hours
Max_Cutting_MTTR              60     Minutes
Max_Sewing_MTTR               40     Minutes
Max_Spreading_MTTR            60     Minutes

OK_Cutting_MTBF               500    Hours
OK_Sewing_MTBF                250    Hours
OK_Spreading_MTBF             500    Hours
OK_Cutting_MTTR               20     Minutes
OK_Sewing_MTTR                15     Minutes
OK_Spreading_MTTR             20     Minutes


LT    → Lead Time
MTBF  → Mean Time Between Failures
MTTR  → Mean Time To Repair
```

Fig. 5.10 Example data from parameter file.

parameters, e.g. existing tolerances and sample sizes may change. The system should be able to handle these changes. These decision variables are not hard coded into the knowledge base, but they are called from the rules as **volunteer data**. In the present system, the values of these decision variables are stored in an ASCII text file. This file is known as the **parameter file** and it can be modified very easily with any line or screen text editor. Some example data from the current parameter file are shown in Fig. 5.10.

The *Weights* are also decision variables, which may need to be modified according to specific evaluation needs. Therefore, the *Weights* are entered in a file known as the **weights file** (Narayanan, 1991). The weights file contains information in an ASCII text file. The information obtained from the apparel enterprise being evaluated is also maintained in a set of ASCII text files that are collectively known as **contractor files**. These files are handled through a 'C' program, which runs the evaluation system by calling Nexpert Object's subroutines. This program calls a subroutine from the action part of the right hand side of a rule for volunteering the enterprise data from the contractor files sequentially.

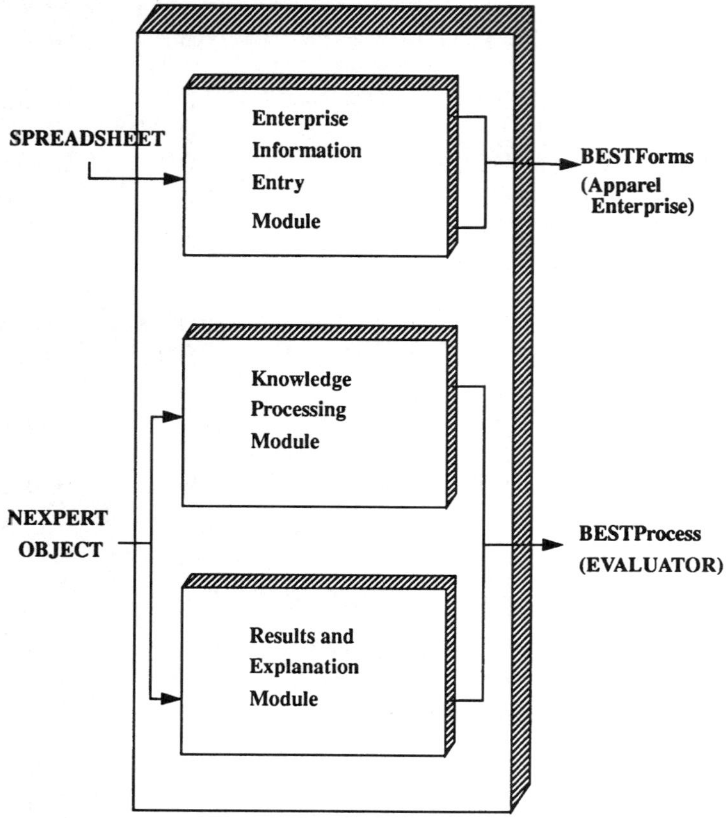

Fig. 5.11 Structure of BEST.

5.9.4 The bid evaluation software tool (BEST)

The implementation of the knowledge framework as a decision support system has resulted in the bid evaluation software tool (BEST). As shown in Figure 5.11, BEST consists of three main modules: the **enterprise information entry** module, the **knowledge processing** module and the **results and explanation** module. It accepts the information about the apparel enterprise being evaluated, processes the information with the help of the knowledge framework (BESTProcess) and provides a summary of the results on the screen. The knowledge processing can also be carried out in **transcript** mode, which provides a step-by-step account of how the *Score* for every object is computed. The results as well as the transcript can be stored in text files for comparing several enterprises. The input to the BEST system (the set of enterprise information files) is generated by a form-based user interface system.

In a decision support system, modularity is extremely important. Therefore, the user interface module is built as a separate, but cooperative module from the knowledge processing modules. This separation of the user interface module and the knowledge processing module would help immensely when BEST is ported to different user interface platforms. Only the enterprise information module would need to be altered while the knowledge processing module would remain the same, thereby facilitating an easy porting process.

5.10 USER INTERFACE FOR BEST

The user interface is one of the major factors that determine the success or failure of a software system like BEST. Therefore, a front-end user interface for obtaining data from the bidder's apparel manufacturing enterprise has been designed for BEST. This front-end will be used by individuals in various departments of the enterprise, who fill in the data in the specified format. Since these users will not necessarily be computer experts, the interface should be simple and user friendly. The interface features of the BEST system's front-end, the enterprise information entry module, are discussed in this section.

5.10.1 User interface requirements of BEST

The BEST system will be used by officers to evaluate enterprises based on the information provided by the participating bidders through the enterprise information entry module. This poses three major design constraints.

1. The data gathering module must be distinctly separable from the evaluating BEST system, though not always separated.
2. The interface should ensure that the data fed to the knowledge processing module of the BEST system are in the format required by BEST.
3. To accommodate a wide variety of users, the enterprise information entry module should be interactive, error corrective, forgiving and friendly.

There are a few additional requirements. For example, not all apparel companies may have computers to enter the data in an electronic format and some companies having computers may prefer to supply the data on paper rather than on a disc. In selecting the user interface, the following three interaction styles were considered:

1. natural language interfaces;

2. menu-based systems;
3. form-filling dialogs.

Based on a comparative assessment of these styles, a form-filling user interface has been selected, since it best meets the design constraints proposed for BEST (Narayanan, 1991). In a form-filling interface, various questions and possible values for the corresponding answers are specified and integrated in a single screen or a logical sequence of screens, that can be scrolled up and down. The user can just fill in the required values in their respective slots. The quality of a form-filling interface depends on three major factors (Gilb, 1975; Gilb and Weinberg, 1977):

1. how well the forms reflect the logic of the system for which the forms serve as the input medium;
2. the clarity of the design and visual presentation of the forms;
3. the integrity of the keyed-in data (correctness and reliability) in various fields, with respect to the program which processes the input data.

In addition, Hayes mentions that a form-filling interface should also support extensive error detection and an integrated on-line help (Hayes, 1985). For BEST, a form-filling interface which satisfies most of these requirements has been designed and developed and is known as BESTForms.

5.11 BESTFORMS IMPLEMENTATION

A widely used implementation vehicle was needed for the development of BESTForms. The implementation vehicle should be a simple tool which can generate the form-based user interface and have provisions to transform the data fed into the forms into a format recognized by the BEST evaluation system. Two options were available to create BEST-Forms:

1. Nexpert Forms – a form-based user interface building tool available with Nexpert Object;
2. a spreadsheet interface, which can either be linked directly to Nexpert Object or which can generate ASCII data from the filled forms.

Nexpert Forms

Nexpert Forms can be directly linked to Nexpert Object, by a set of command files known as run time definition (RTD) files. These RTD files connect the data input locations in the forms to the corresponding data in Nexpert Knowledge Processing.

One of the major requirements of a good user interface is that user actions should be reversible. In Nexpert Forms, once one form has been completed and the next form is shown, it is not possible to go back to the previous form to make any corrections. Therefore, a mistake made by the user on one form needs to be corrected while still on that form, and before proceeding to fill in the next form. Otherwise, the user is forced to restart the session. This type of interaction would be in violation of the principle of reversibility of user actions (Baecker and Buxton, 1987). Therefore, Nexpert Forms did not prove to be the ideal user interface development tool for the enterprise information entry module.

Spreadsheet templates

In a spreadsheet, all the elements of Nexpert Forms can be easily created. Moreover, errors can be corrected easily without any excessive user interaction. Hence the spreadsheet user interface was deemed to be an appropriate one for the system. Quattro has been chosen as the spreadsheet package, because Quattro can accept templates from other common spreadsheet packages such as Lotus 1–2–3 and Microsoft Excel. These templates together with their built-in programs constitute BEST-Forms, the front-end user interface to BESTProcess.

5.11.1 Features of BESTForms

The BEST system requires more than 500 information entities. Hence a logical separation of the input forms into groups of forms becomes necessary. A grouping based on the BEST class hierarchies would be conceptually clean and easy from the evaluator's point of view, but it would not offer any benefit to the people using BESTForms, i.e., the company personnel providing the enterprise information. Hence the grouping should be aimed at simplifying the data-gathering process. The grouping of all data items pertaining to a specific department on a stand-alone form for that department would be ideal from the enterprise's point of view. Each department in the apparel manufacturing facility can then enter data on the corresponding departmental form. For these reasons, five different forms have been designed:

- overall bidder information form;
- spreading and cutting room form;
- sewing room form;
- quality control form;
- maintenance form.

These five forms are known as the **departmental** forms. There is also a sixth form known as the **master information** form which consolidates the data entered on the five departmental forms. This master information form is intended for use by the evaluator. On this form the evaluator can enter

SPREADING AND CUTTING ROOM FORM

Floor and Machine Dimensions

Spreading & Cutting Floor Space (sq ft):	________
Average Length of Spreading Tables (ft):	________
Average Width of Spreading Tables (ft):	________
Average Length of Cutting Tables (ft):	________
Average Width of Cutting Tables (ft):	________

Number of Machines

Total Number of Spreading Tables:	________
Total Number of Cutting Tables/Machines:	________

Do the Spreading Machines have these features?

	(Y/N or T/F)
Automatic Tensioning	________
Turn Table	________
Ply Counter	________
End Catcher	________
Surface Leveling	________
Cloth Inspection Light	________
Bolt Drive	________
Stop Control	________
Defect Sensor	________
Ply Cutter	________
Auto Ply Alignment	________

Fig. 5.12 Example departmental form.

the past quality performance score for the enterprise being evaluated. All the departmental forms have built-in programs known as **spreadsheet macros**, which check the data and convert it into ASCII data files in the format required by the BESTProcess system. All these departmental forms are also available on paper for bidders not using the electronic version (Fig. 5.12 shows part of the spreading and cutting room form).

5.11.2 Data validation in BESTForms

The macros check the validity of the data entered in the forms. Error handling depends on the type of the erroneous datum. Three types of data are sought on the forms:

1. essential data without default values;
2. essential data with default values;
3. optional data.

If an essential datum without default value is missing or a wrong type is entered, the system cannot function any further and the user is prompted to enter a value. When the user enters an erroneous (or no) value, the system assigns the default value to that slot but the user is still given the option to alter it. These default values would result in the worst score for the factor to which the specific data items contribute. Hence it is better for the bidder to enter the datum rather than rely on the system to assign its default value. For the optional data, the user can respond with 'unknown', but here too it may lead to the worst score for the corresponding higher level factor.

If any datum is entered incorrectly, BESTForms displays an error message and pinpoints the error to the user by moving the cursor to the spreadsheet cell where the error occurred. The user is given the option either to correct only that specific datum at the error prompt itself and continue checking, or go back to the spreadsheet cell for additional corrections or data entry. Another important feature of the interface is that while checking, the system can interpret any uniquely identifiable set of characters and replace the set of one or more characters by the complete required data value. For example, for any question requiring a boolean answer, the response 'y' or 't' is interpreted as 'True' and 'n' or 'f' is interpreted as 'False' and '?' is interpreted as 'Not Known'. Another example relates to the question about the type of lint cleaning system, where 'b' is interpreted as 'blower only', 's' is interpreted as 'suction' and 'bs' is interpreted as 'blower and suction' type. This uniquely identifiable set of minimum number characters makes it very easy for the user to enter data, because most of the data entry could be carried out in a single keystroke.

5.11.3 Navigation facilities in BESTForms

Every screen in BESTForms contains navigation instructions listed at the corners or bottom of the screen. The user can follow these instructions to enter data in the entire template and finally check the data and create the data file. This data file, which can be used by the BEST system for processing, is created only when all the entered data are in the correct format. The interface screens are color-coded to enable the user to easily identify the data entry locations. Also, the template is protected in such a way that data can be entered only in the data entry cells. The system displays an error message when a modification or data entry is attempted in any of the protected cells. The macros and other data cells which do not typically concern the user are hidden and are invoked only when the user checks or prints the data file. An **on-line help** facility is available on all the forms, and this can be invoked at any time during data entry. All these actions are carried out with the help of the macros built into each of the spreadsheet templates.

5.12 BEST RESULTS

The result of BESTProcess is a set of scores on a 0–4 scale for all the objects identified in AEEF as factors determining the capability of the enterprise. The system provides the evaluator with a brief summary of the results. Once the evaluation is complete, the system shows a result screen which contains the *Overall_Score* for the enterprise evaluated (Fig. 5.13). The result screen also shows the breakdown of the score to the next two levels of factors under *Overall_Score*, along with their respective weights. The results can be stored in a text file. As mentioned earlier, the evaluator can choose to go through the complete evaluation process in the transcript mode to create a step-by-step account of the process and store it in a text file.

The BEST system has been tested and debugged using assumed data leading to extreme scores, as well as assumed data with realistic values. Testing with actual enterprise data from major apparel manufacturers and comparison the results with human evaluators have also been carried out. BEST results match the evaluations by industry experts thus validating the system's knowledge and inference process. BEST takes approximately three minutes to evaluate a bid. This is in contrast to the several weeks spent when a bid is manually evaluated.

5.13 CONCLUSIONS

Most of the current vendor evaluation programs in the industry are based on past performance criteria rather than on the current manufacturing capability of the vendors. There is an absence of knowledge-based vendor rating decision support in the apparel industry. A set of factors which affects the manufacturing capabilities of an enterprise has been identified through the knowledge acquisition process. The effects of these factors on the overall possibility of getting a quality product at the right time from an enterprise has also been estimated. All the major enterprise capability factors have been abstracted and grouped into quality, production and financial capabilities.

BID EVALUATION SOFTWARE TOOL (BEST) RESULTS)

Date: Aug 23 1991

Bidder: Enterprise C **Overall Score:** 2.48

Bid Value: $195000 **Order Size:** 93600 **Trousers/Day:** 1440

Walsh-Healey Category: Manufacturer **Contract Type:** SBSA

Distribution of Overall Score

Criteria	Weight	Score (0 to 4 Scale)
Quality Capability	0.45	2.58
Production Capability	0.35	1.95
Financial Capability	0.20	3.20
Total	1.00	2.48

Distribution of Quality Capability

Criteria	Weight	Score (0 to 4 Scale)
QC Practices	0.27	3.71
QC Human Resources	0.18	2.13
Maintenance	0.06	2.51
QC Info System	0.10	2.40
QC Organization	0.10	2.00
Manufacturing Features	0.27	1.96
Past Quality	0.02	3.84
Total	1.00	2.58

Fig. 5.13 Sample BEST results.

The knowledge-based framework (apparel enterprise evaluation framework – AEEF) for evaluating utility trouser manufacturing enterprises has been developed. The AEEF has been implemented as a decision support system (bid evaluation software tool – BEST), which can be used to evaluate competing apparel manufacturing enterprises to perform on a contract. BEST is implemented with the hybrid object-oriented and rule-based expert system shell Nexpert Object. The knowledge in AEEF has been structured hierarchically and represented using classes and objects, and the inference mechanism is implemented as a complex set of rules in Nexpert Object. There are three separate modules in BEST: enterprise information entry module, knowledge processing module and results and explanation module. These modules can function independent of one another, but can also be connected to construct an integrated decision support system. A form-based user interface (BEST-Forms) has also been created to obtain the information from apparel manufacturing enterprises. BEST has been tested and validated with data from apparel plants.

BEST represents the first application of knowledge-based systems technology for evaluating the capability of an apparel manufacturing enterprise to perform on a contract and fills a long-felt need for such a tool in the apparel industry. Though the knowledge incorporated in BEST is specific to the domain of utility trouser manufacturing, its methodology can be extended to other domains (food, medicine, automobile parts etc.) also. While BEST is specifically suited to DoD procurement policies, its underlying framework has been developed in a modular fashion so that it can be extended to suit other organizations including commercial apparel manufacturers. Hence the apparel industry, as a whole, can stand to benefit from this framework.

ACKNOWLEDGEMENTS

This research was funded by the US Defense Logistics Agency under Award DLA-900-87-D-0018/0002. The authors would like to thank officials from DLA, the Defense Personnel Support Center, the field offices of DLA in Marietta and Warner Robbins, GA, and Birmingham, AL, for their input and participation during the course of the research. Dr Howard Olson deserves sincere appreciation and thanks for his input and comments during the later stages of system development. Thanks are also due to Mr Dale Stewart, Research Technician, for his assistance in developing the machine classification scheme. Input from fellow researchers in the lab was helpful in finalizing the evaluation framework. Finally, the authors would like to acknowledge the assistance of the American Apparel Manufacturers Association and their member companies during the course of the research.

REFERENCES

Apparel (1989) *Apparel Plant Wages Survey*, American Apparel Manufacturers Association Inc. Arlington, VA.

ASL (1985) *Acceptable Suppliers List (ASL) No. 63*, Defense Support Personnel Center, Philadelphia, PA.

Baecker, R.M. and Buxton, W.A.S. (1987) *Readings in Human–Computer Interaction*, Morgan Kaufmann, San Mateo, CA, pp. 468–70.

Barnaby, J.A. and Bohannon, K.J. *(1975) An investigation of negative pre-award surveys as an indicator of a contractor's inability to meet a delivery schedule.* MS Thesis, School of Systems and Logistics, Air Force Institute of Technology, Wright-Patterson AFB, OH.

Best Choice 3 (1990) *User Manual*, Sterling Castle Inc., Marina del Ray, CA.

Cormany, T.P. and Donnellan, J.P. (1975) *Criteria for evaluation of contractor management potential during the source selection process for the acquisition of major weapons systems.* MS Thesis, School of Systems and Logistics, Air Force Institute of Technology, Wright-Patterson AFB, OH.

DDD (1987) *DDD-L-20E. Federal Specification for Label: Clothing, Equipage and Tentage*, September 28 US Government Printing Office, Washington, DC, p. 22.

DPSC (1987) *DPSC Handbook for Formal Source Selection (Draft)*, Defense Personnel Support Center, Philadelphia, PA.

DPSC (1988) *DPSC Memorandum to Prospective Clothing. Textile Equipment and Footwear Bidders*, Defense Personnel Support Center, Philadelphia, PA.

Eberly, N. (1990) Cutting room update, part 2. *Apparel Manufacturer*, **7**, 30–38.

Edwards, V.J. (1989) *Source Selection for Defense Logistics Agency, 2nd edn*, Educational Services Institute, Falls Church, VA.

Gilb, T. (1975) Laws of unreliability. *Datamation*, **21**(3), 81–85.

Gilb, T. and Weinberg, G.M. (1977) *Humanized Input: Techniques for Reliable Keyed Input*, Winthrop Publishers, Cambridge, MA.

Green, P.E. and Carmone, F.J. (1970) *Multidimensional Scaling and Related Techniques in Marketing Analysis*, Allyn and Bacon, Boston, MA.

Green, P.E. and Rao, V. (1972) *Applied Multidimensional Scaling: A Comparison of Approaches and Algorithms*, Holt, Rinehart and Winston, New York.

Green, P.E. and Wind, Y. (1973) *Multiattribute Decisions in Marketing*, Dryden Press, Hillsdale, IL.

Hayes, P. (1985) Executable interface definitions using form-based interface abstractions, in Advances in Human–Computer Interaction (ed. E.R. Hartson), Ablex Publishing Corp., Norwood, NJ.

Hodgins, H.A. (1990) From edge guide to electronics. *Apparel Manufacturer*, **2**(3), 42–43.

Hudson, P.B. *(1988) Guide to Apparel Manufacturing*. MEDIApparel Inc., Greensboro, NC.

Hwang, C.L. and Yoon K. (1981) *Multiple Attribute Decision Making*, Springer-Verlag, New York, pp. 41–47.

Jayaraman, S. (1989a) On a manufacturing enterprise architecture, in IJCAI 1989 – *Workshop on Integrated Architectures for Manufacturing*, Detroit, MI.

Jayaraman, S. (1989b) Design and development of a knowledge-based framework for trouser procurement. *Internal Technical Report SJ-TR-PROC-8901*, Georgia Institute of Technology, Atlanta, GA.

Jayaraman, S. (1990) Design and development of an architecture for computer-integrated manufacturing in the apparel industry, part I: basic concepts and methodology selection. *Textile Research Journal*, **60**(5), 247–54.

Jones, C. (1990) Determining cutting space requirements. *Apparel Manufacturer*, **2**(3), 26–29.

Lange, G. and Heuermann, R. (1973) *An Analysis of the Army's Contractor Performance Evaluation Program*. US Army Logistics Management Center, Institute of Logistics Research, Fort Lee, VA.

Liu, T.S., Narayanan, S., Subramanian, V. and Konz, S. (1990) *Relative vs. absolute rating, in Proceedings of the Human Factors Society 34th Annual Meeting,* Orlando, FL, pp. 1229–31.

McLennen, R.K. *(1984) The feasibility of a decision support system for the evaluation of source selection evaluation criteria.* MS Thesis, Air Force Institute of Technology, Wright-Patterson AFB, OH.

MIL (1964) *MIL-STD-105D. Military Standard Sampling Procedures and Tables for Inspection by Attributes,* March 20, US Department of Defense, Washington, DC.

MIL (1984a) *MIL-C-24915 Military Specification for Cotton Denim Cloth,* June 6, US Department of Defense, Washington, DC.

MIL (1984b) *MIL-T-87062A. Military Specification – Men's Utility Denim Trousers,* June 11, US Department of Defense, Washington, DC.

MIL (1987) MIL-STD-1488F. *Military Standard for Evaluating Quality of Trousers,* July 31, US Department of Defence, Washington, DC.

Narayanan, S. (1991) *A knowledge-based framework for apparel enterprise evaluation.* Masters Thesis, Georgia Institute of Technology, Atlanta, GA.

Narayanan, S., Olson, L.H. and Jayaraman, S. (1994) AEEF – a knowledge-based framework for apparel enterprise evaluation. *Expert Systems with Applications,* **7**(2), 337–56.

Nexpert Object (1988a) *User Manual,* Neuron Data Inc., Palo Alto, CA.

Nexpert Object (1988b) *Programmer Manual,* Neuron Data Inc., Palo Alto, CA.

Pingel, R.D. (1981). *Proposed system for the use of evaluation factors in the source selection of service contractors.* MS, Thesis, Naval Postgraduate School, Monterey, CA.

Rothman, P. (1989) Selecting an uncertainty management system. *AI Expert,* **7**, 56–62.

Schuman, J.D. and Vitelli, J. (1978). *A test to evaluate a proposed Air Force logistics command indicator of contractor performance.* MS Thesis, School of Systems and Logistics, Air Force Institute of Technology, Wright-Patterson AFB, OH.

Shafer, G. *(1976) A Mathematical Theory of Evidence,* Princeton University Press, Princeton, NJ.

Solinger, J. (1980) *Apparel Manufacturing Handbook.* Von Nostrand Reinhold, New York.

Srinivasan, K. (1991) *FDAS: a knowledge-based framework for analyzing defects in woven structures.* Masters Thesis, Georgia Institute of Technology, Atlanta, GA.

Srinivasan, K., Dastoor P.H. and Jayaraman, S. (1992) FDAS: architecture and implementation. *Expert Systems,* **9**(3), 115–24.

V-B (1984) *V-B-871E. Federal Specification for Button, Sewing Hole and Button Staple,* March 28 US Government Printing Office, Washington, DC.

V-F (1987) *V-F-106F. Federal Specication for Fastener, Slide and Interlocking,* June 23, US Government Printing Office, Washington, DC.

V-T (1982) *V-T-280C. Federal Specification for Thread, Gimp, Cotton, Button Hole,* June 28, US Government Printing Office, Washington, DC.

Zadeh, L.A. (1965) Fuzzy sets. *Information and Control,* **8**, 338–53.

CHAPTER SIX

Heuristic decision support system database structure for diagnostic expert systems

John W. Peters and Adedeji B. Badiru

6.1 INTRODUCTION

Expert systems have been developed for practical applications in areas such as oil exploration, chemical structure analysis, medical diagnosis, industrial equipment troubleshooting, personnel training, farm management, business credit analysis (Arend, 1989), mathematical analysis, configuration of computer equipment, production scheduling and control, cancer classification, space vehicle electric power management, monitoring of biological activities in ponds and assisting nurses in scheduling long-term rehabilitation health care programs (Blackman, 1990; Bowerman and Glover, 1988; Buchanan and Shortliff, 1984; Codd, 1990; Evans, 1988; Kerklof, 1989; Myers, 1988). Knowledge and data engineering for expert systems requires the integration of theoretical and heuristic approaches. Diagnostic expert systems are used in many industrial settings (Gorman, 1987; Harmon and King, 1985; Hashemi-Nassab, 1989; Mahler, 1989; Mitchell, 1988). They are used to interpret alarm conditions in automated manufacturing plants and to determine proper corrective action (Herrod, 1988). Effective use of these diagnostic tools requires that the knowledge of experts be coupled with theoretical data characteristics to achieve a reasonable representation of the problem scenario.

Currently, expert systems are developed for a machine process, or class of machines. They are generally prepared by a knowledge engineer who has considerable experience with computers and expert systems. The knowledge base is primarily derived from information which is supplied by human experts and interpreted by the knowledge engineer. Since some of these systems may cost as much as a quarter of a million dollars (Boult, 1988) and may take a year or more to develop, they can only be considered for long life, very expensive machinery, or by a vendor as a maintenance aid to customers where the vendor has many units operating in similar conditions.

In some industries, for example, an operation may have many types of machines, and probably only a few of any one make or model. The conditions in which the equipment operates vary greatly from one operation to another. These make problem diagnosis very difficult. Rowan (1989) lists the following criteria as requirements for constructing a diagnostic system:

1. steps in a preferred sequence;
2. audit trail or journal of steps;
3. feedback to engineering and management;
4. straightforward system maintenance;
5. expandability without major rewrite.

Practical diagnostic expert systems require the integration of theoretical and heuristic approaches. Theoretical data analysis will likely be impractical for equipment diagnosis in large and dynamic maintenance operations. To achieve practicality for most real-world problems, the heuristic expertise of humans must be incorporated into expert system knowledge bases.

This chapter presents a heuristic database structure for diagnostic expert systems in manufacturing. The approach is based on the heuristic matrix methodology developed by Peters (1992; Raiffa, 1968). The database structure facilitates the handling of data uncertainty in diagnostic expert systems. The proposed methodology permits the utilization of data which is resident or could be entered, retained and retrieved from a company's MIS databases.

The matrix consists of a list of problems or symptoms and a list of causes. Each symptom can point to many causes, and each cause can reflect several symptoms which relate to it. The proposed approach will allow field personnel, knowledgeable in the problem being diagnosed, to use information available to them in maintenance records to develop a diagnostic expert system for equipment. The approach uses frequency of occurrence of the problems or symptoms of failures or malfunctions correlated with the determined causes to induce IF/THEN rules. These rules will utilize a heuristic certainty indicator to rank the possible causes in a hierarchy ranging from most likely to least likely. A complete detail of the heuristic matrix approach is presented by Peters (1992).

6.2 HANDLING UNCERTAINTY IN DIAGNOSTIC SYSTEMS

Many expert system developers use various forms of Bayesian logic to manage uncertainty or to ascribe levels of certainty to rule conclusions. Two examples of this are Thompson and Thompson's (1986) use of a communication theory called 'entropy' and Bayesian belief networks. Entropy is a measure of the uncertainty of the classification of an object. Entropy ($H(C)$) can be computed as follows:

$$H(C) = -\sum_{i=1}^{N} P(c_i)\log_2 P(c_i) \tag{6.1}$$

where N is the number of classes and $P(c_i)$ is the probability of an object being in class i. $H(C)$ represents the entropy of all N classifications of an object. Each successive level of calculation is dependent on the previous, thus:

$$H(C|a_j) = -\sum_{i=1}^{N} P(c_i|a_j)\log_2 P(c_i|a_j) \tag{6.2}$$

where $H(C|a_j)$ is the entropy of class C given a_j and a_j is the jth value of attribute A of the object under consideration. The total entropy of attribute A is the weighted average of each value of the attribute $(H(C|a_j))$:

$$H(C|A) = -\sum_{j=1}^{N} p(a_j)H(C|a_j) \tag{6.3}$$

where $p(a_j)$ is the probability of the jth value of attribute A. The higher the total entropy $(H(C|A))$, the more uncertain the identification of the attribute. Thus, more decision emphasis will be placed on attributes with lower entropy. This premise is the converse of Bayesian belief networks, which estimate the certainty that the rule is true, rather than the uncertainty (Morawski, 1989a). Bayesian methods consider certainty in terms of either belief or disbelief. Another factor that influences human reasoning is a measure of lack of understanding, not of whether to believe or disbelieve, but of ignorance or vagueness related to the decision premise. Two methods of managing uncertainty that attempt to recognize this believe/disbelieve notion are the Dempster–Shafer theory and fuzzy logic.

6.3 DEMPSTER–SHAFER THEORY

The Dempster–Shafer (D–S) theory is a combination of Dempster's rules for combining belief functions with Shafer's representation of beliefs. Dempster's rule makes possible the combining of probabilities based upon different sources of information (Falkenhainer, 1986). This is accomplished by introducing a third value into the set besides the different possible hypotheses to account for lack of knowledge or 'ignorance'. Shafer's method of representation allows for the representation by intervals (Falkenhainer, 1986) or subsets (Sumita *et al.*, 1989). Thus, there are three main differences between the Dempster–Shafer theory and Bayesian logic:

1. D–S accepts an incomplete probabilistic model.
2. D–S accepts contradictory or incompatible propositions because it interprets strength of evidence, not as likelihood but as surface evidence of truths.
3. D–S does not have to provide full answers to probabilistic queries, it will provide partial answers (Pearl, 1988).

As in Bayesian belief networks, the choice between hypotheses is made by comparing the belief values assigned to each (Schank and Abelson, 1977). Because of the differences described above there is no single value for the belief for any individual hypothesis. The uncertainty function, U (Badiru, 1992) is another function that helps us understand the 'strength of relationship' between hypotheses. Assume a set $\boldsymbol{H}$ consists of three mutually exclusive hypotheses presented below (Hau and Kashyap, 1989): $\mathbf{H} = \{h_1, h_2, h_3\}$.

Suppose two independent knowledge sources have provided two independent premises: $m_1 = \{h_1, h_2, h_3\}$ and $m_2 = \{h_1, h_2, h_3\}$, such that $P(m_1) + P(m_2) \leq 1$. Assume the Bayesian probabilities based on the knowledge mentioned above are:

$$m_1(\{h_1\}) = 0.75 \quad m_2(\{h_1\}) = 0.05 \quad m_1(\{h_2\}) = 0.20$$
$$m_2(\{h_2\}) = 0.20 \quad m_1(\{h_3\}) = 0.05 \quad m_2(\{h_3\}) = 0.75$$

where $m_i(\{h_j\})$ is the probability of h_j given that premise m_i is true. Using Dempster's rule of combination, the belief in hypothesis h_j can be calculated as:

$$Bel(h_j) = m_1 \odot m_2 = \frac{\prod_{i=1}^{M} m_i(\{h_j\})}{\sum_{k=1}^{N}\left(\prod_{i=1}^{M} m_i(\{h_k\})\right)} \tag{6.4}$$

Thus,

$$Bel(h_2) = \frac{0.20(0.20)}{0.75(0.05) + 0.20(0.20) + 0.05(0.75)} = 0.35 \tag{6.5}$$

The belief that the h_j is not correct, Bel(h_j^c), can be similarly calculated and plausibility of hypothesis Pl(h_j) then becomes:

$$Pl(h_j) = 1 - Bel(h^c) \tag{6.7}$$

and uncertainty is defined as:

$$U(h_j^c) = Pl(h_j^c) - Bel(h_j^c) \tag{6.8}$$

Table 6.1 Dempster–Shafer function values

H	m_1	m_2	$Bel(h_j)$	$Bel(h_j^c)$	Pl	U
h_1	0.75	0.05	0.33	0.21	0.79	0.46
h_2	0.20	0.20	0.35	0.43	0.57	0.22
h_3	0.05	0.75	0.33	0.21	0.79	0.46

The Dempster–Shafer function values for all the functions for the three hypotheses are presented in Table 6.1. The values in Table 6.1 provide a measure for the belief and uncertainty of a premise. The normalization procedure utilized in Dempster's rule of combination tends to place greater emphasis on consensus among knowledge functions than on high level of confidence displayed by one knowledge function which is not shared by others. As decision criteria are analyzed, belief level can be varied between 'Bel' and 'Pl'. In the example, one would put the greatest faith in hypothesis h_2 since h_1 and h_3 exhibit higher potential upper bounds of belief ($\text{Pl}(h_1) = \text{Pl}(h_3) = 0.79$ and $\text{Pl}(h_2) = 0.57$). The tendency would be to disregard h_1 and h_3 unless additional knowledge was available because of the higher degree of uncertainty U. This also presents an opportunity for displaying great belief in a hypothesis which is only weakly supported by the knowledge functions. For example if Bayesian probabilities in the case above were changed as follows:

$$m_1(\{h_1\}) = 0.99 \quad m_2(\{h_1\}) = 0.00 \quad m_1(\{h_2\}) = 0.01$$
$$m_2(\{h_2\}) = 0.01 \quad m_1(\{h_3\}) = 0.00 \quad m_2(\{h_3\}) = 0.99$$

Then, the belief in h_2 becomes:

$$Bel(h_2) = \frac{0.01(0.01)}{0.99(0.00) + 0.01(0.01) + 0.00(0.99)} = 1.00 \tag{6.9}$$

$$Bel(h_1) = Bel(h_3) = Bel(h_3^c) = 0.0 \tag{6.10}$$

6.4 FUZZY LOGIC

Fuzzy logic is a heuristic method of easily transforming verbal expressions into numerical values. Fuzzy sets are essentially two sets of variables, the value of which can vary continuously from 0 to 1 and describe two different related aspects of the antecedent–consequent relationship. These values can be described in a manner such that X is a measure of how frequent or significant an event is when Y is present, or Y is the degree of confirmation or importance given a value of X (Adlassnig, 1986, Wilhelm and Parsaei, 1991; Zadeh, 1992). Adlassnig (1986), in his medical diagnostic example, defines the two parameters as:

- frequency of occurrence (o);
- strength of confirmation (c).

The same could be applied to an equipment diagnostic example. In this case, the general form of the rule would be: IF premise THEN conclusion WITH (o, c). Example:

How frequently a person gets sick (o)
How sick the person is each time he gets sick (c)
IF Jack is **sick** THEN he is **moderately sick**.

Fuzzy linguistic values for o and c can be acquired in one of two ways: Information acquired from experts or statistical inference from databases. If the values are determined statistically, the following formulas are applied (Adlassnig, 1986):

$$u^o = \frac{F(P \cap C)}{F(C)} = F(P|C) \tag{6.11}$$

$$u^c = \frac{F(P \cap C)}{F(P)} = F(C|P) \tag{6.12}$$

where

- u^o = frequency of occurrence;
- u^c = strength of confirmation;
- P = the observed problem;
- C = the determined cause;
- $F(P \cap C)$ = the absolute frequency of occurrence of P and C;
- $F(C)$ = the absolute frequency of occurrence of C;
- $F(P)$ = the absolute frequency of occurrence of P;
- $F(P \mid C)$ = conditional frequency of P given C;
- $F(C \mid P)$ = conditional frequency of C given P.

Table 6.2 shows a hypothetical example of how the knowledge might be displayed. An example of a rule that demonstrates this relationship for an automobile transmission diagnostic system would be:

IF engine starts with transmission in gear
THEN misadjusted or defective neutral start switch
WITH (u^o = Often = 0.90, u^c = Weak = 0.25)

Table 6.2 Numeric values for fuzzy linguistic values

	Strength of confirmation (u^o)				
Frequency of occurrence (u^o)	*Very strong*	*Strong*	*Medium*	*Weak*	*Very weak*
Always	1.00/1.00	1.00/0.75	1.00/0.50	1.00/0.25	1.00/0.00
Often	0.90/1.00	0.90/0.75	0.90/0.50	0.90/0.25	0.90/0.00
Normal	0.50/1.00	0.50/0.75	0.50/0.50	0.50/0.25	0.50/0.00
Seldom	0.10/1.00	0.10/0.75	0.10/0.50	0.10/0.25	0.10/0.00
Never	0.00/1.00	0.00/0.75	0.00/0.50	0.00/0.25	0.00/0.00

6.5 PROBLEM–CAUSE RELATIONSHIPS

It is essential to define some basic parameters of the problem and cause relationships of the diagnostic decision process.

1. A symptom or problem may point to several causes.
2. Many diagnoses are made because more than one symptom or problem is recognized that points to a certain cause.
3. Some symptoms or problems may be more easily recognized than others and also may be more easily recognized when a particular cause is actually present.
4. When reporting that certain symptoms or problems exist the operator may:
   - correctly identify the problems;
   - commit an error by selecting an invalid problem;
   - commit an error by omitting a valid problem.
5. A cause may exist that has not been previously defined.
6. A symptom or problem may appear that has not been observed.
7. More than one cause may be present that cannot be readily distinguished.
8. The information given may not be sufficient to permit a clear choice of the cause.

A diagnostic expert system must recognize and address these relationships and be able to reduce them to a manageable set of rules or knowledge bits. It will be shown that the proposed decision rule matrix provides a simple method of making this transition. A typical matrix will be used to demonstrate why a probabilistic approach is not practical and how a heuristic approach can simplify the solution.

6.6 DESIGN OF THE MATRIX STRUCTURE

The sample matrix shown in Table 6.3 will be used to demonstrate the new concept. The matrix represents a hypothetical situation in which three recognizable problems (P_1, P_2, P_3) have been found to be indicative of four known causes (C_1, C_2, C_3, C_4). The value of N represents the number of occurrences of each condition that has been observed in the past. This information can be derived from history available in a corporate database (Kaplan, 1984; Merino, 1989). The process of extracting and making use of this knowledge from the databases is demonstrated in this chapter. From the basic parameter of the problem–cause relationship described previously, we can make the following inferences about the values in the matrix:

- $N(P_i)$ = the number of times that problem i is noted;
- $N(C_j)$ = the number of times that cause j has been the nature of the malfunction that required repair;
- $N(P_i, C_j)$ = the number of times when P_i was indicated as the problem and C_j as the cause.

Table 6.3 Decision rule matrix

	P_1 $N(P)_1$	P_2 $N(P_2)$	P_3 $N(P_3)$	
C_1	$N(P_1, C_1)$	$N(P_2, C_1)$	$N(P_3, C_1)$	$N(C_1)$
C_2	$N(P_1, C_2)$	$N(P_2, C_2)$	$N(P_3, C_2)$	$N(C_2)$
C_3	$N(P_1, C_3)$	$N(P_2, C_3)$	$N(P_3, C_3)$	$N(C_3)$
C_4	$N(P_1, C_4)$	$N(P_2, C_4)$	$N(P_3, C_4)$	$N(C_4)$

One or more problems must be found to instigate the search for a cause. These problems must be either the known problems or some previously unknown problem. Therefore,

$$\sum_{i=1}^{3} N(P_i, C_j) \geq N(C_j), j = 1, 2, 3, 4 \tag{6.13}$$

Also, each time a problem is identified, the cause must be one of the identified causes or some not previously identified causes or a combination of causes. Thus,

$$\sum_{j=1}^{4} N(P_i, C_j) \leq N(P_i), i = 1, 2, 3 \tag{6.14}$$

6.6.1 Combinatorial size of the data space

Figure 6.1 presents problem–cause relationships. The entire matrix could be expanded as shown in Table 6.4 to provide a probabilistic solution with a simple rule for each combination of problem and cause, such as:

IF P_i THEN C_j with (certainty level).

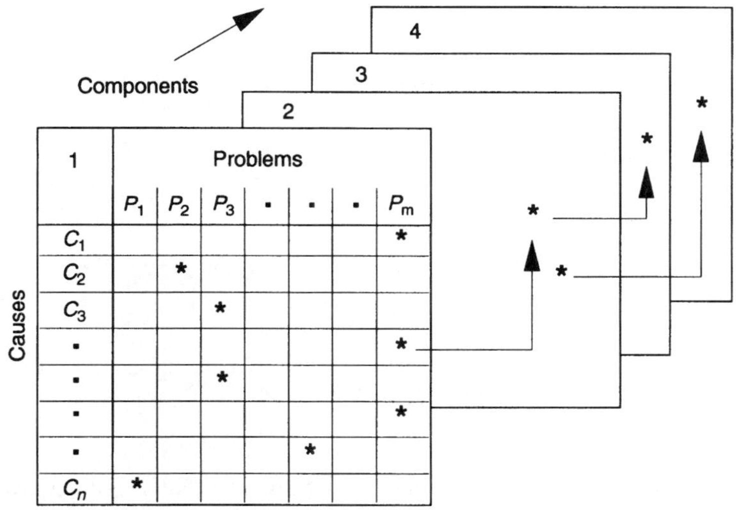

Fig. 6.1 Problem–cause relationships.

Although this method of solution seems straightforward, the expanded matrix has gone from the original 12 cells for which a frequency of occurrence must be developed to 128 cells. The diagnoses could then be solved with something less than 128 rules. No rules would be required where there was no likelihood that problem P_i was symptomatic of C_j. The data from the databases from which we are getting the information are sketchy and contain human error. For example, an operator may easily omit reporting one of several problems observed, or may report a problem that has not actually occurred. With the potential for these errors, it is doubtful that any strong degree of reliability could be placed on the 'sureness' factor for cells in which the occurrence was infrequent.

These problems would be compounded if the original matrix was of the order of 10–15 by 15–20. This condition is not unlikely in an actual diagnostic system. In this case, the number of rules required for the expanded case could be from $2^{10} \times 2^{15}$ to $2^{15} \times 2^{20}$ (from 33 million to 34 billion). It can be seen that this form of solution rapidly becomes

Table 6.4 Expanded matrix for probabilistic solution[a]

	P_1	P_2	P_3	P_1, P_2	P_1, P_3	P_2, P_3	P_1, P_2, P_3	P^c
C_1								
C_2								
C_3								
C_4								
C_1, C_2								
C_1, C_3								
C_1, C_4								
C_2, C_3								
⋮								
C_1, C_2, C_4								
⋮								
C_1, C_2, C_3, C_4								
Cc								

[a] C_j = known cause j; P_i = known problem i; P^c = previously unrecorded (unknown) problem; C^c = previously unrecorded (unknown) cause.

impractical. The number of rules will be extremely large regardless of which of the above methods of dealing with uncertainty is used. Also, the mathematical concepts of these methods for dealing with uncertainty are complex and, therefore, very difficult for day-to-day equipment diagnosis.

6.7 EXAMPLE OF THE MATRIX

The construction of the heuristic matrix is discussed in this section. The matrix presents a unique method for determining the relative certainty associated with determining the cause or causes of a malfunction (Peters, 1992). It can be applied to the diagnosis of any equipment. It presents an alternative to the use of Bayesian logic, the Dempster–Shafer theory, fuzzy sets or other previously applied methods. The matrix is also a method by which information contained in operating and maintenance databases can be used to determine the certainty of a diagnostic problem–cause relationship. Rules can be generated from these calculations. An individual matrix will be used to represent the following:

- a total system or machine, in which case the solution of the matrix may simply identify the most likely affected subsystem or major component;

- a subsystem or component, which may result in identifying the most likely cause of the problem, or may simply further narrow the search field;
- the lowest level subsystem or component, the point where the specific cause is identified.

The best way to illustrate the methodology of the heuristic matrix calculation is by using an example. The three by four matrix presented earlier will be used for the illustration. Hypothetical numbers of observations for each case have been added to allow the use of numerical sample calculations of the example, as well as developing equations to simplify the explanation. An additional row has been added to show the number of times that the problem named was not correctly associated with the cause that was determined. This is presented in Table 6.5, where:

- $N(P_i)$ = the number of times that problem i is noted;
- $N(C_i)$ = the number of times that cause j has been the nature of the malfunction that required repair;
- $N(P_i, C_j)$ = the number of times when P_i was indicated as the problem and C_j as the cause.

Table 6.5 Decision rule matrix with hypothetical values

	P_1 $N(P_1) = 40$	P_2 $N(P_2) = 60$	P_3 $N(P_3) = 70$	
C_1	$N(P_1, C_1) = 0$	$N(P_2, C_1) = 16$	$N(P_3, C_1) = 28$	$N(C_1) = 30$
C_2	$N(P_1, C_2) = 12$	$N(P_3, C_2) = 0$	$N(P_3, C_2) = 15$	$N(C_2) = 20$
C_3	$N(P_1, C_3) = 28$	$N(P_2, C_3) = 32$	$N(P_3, C_3) = 25$	$N(C_3) = 40$
C_4	$N(P_1, C_4) = 0$	$N(P_2, C_4) = 10$	$N(P_3, C_4) = 0$	$N(C_4) = 10$
Incorrect C	$N(P_1, \text{not } C) = 0$	$N(P_2, \text{not } C) = 2$	$N(P_3, \text{not } C) = 2$	

The assumption used in the matrix is that the numbers are the observed results based on the analysis of historical data from 100 logged failures of the equipment. These numbers are the type of information which one would reasonably expect to be obtainable from well-organized operating and maintenance data records if a proper procedure were in place to accumulate it. The nature of the information and how the database can be organized are discussed by Peters (1992).

6.7.1 Matrix evaluation of problem–cause relationships

A list of basic parameters of problem–cause relationships in diagnostic expert systems was presented earlier. The following paragraphs discuss how the matrix deals with each of the parameters.

(a) A symptom or problem may point to several causes

Each cause suggested by a symptom can be identified by reviewing the matrix. The matrix shows how often the problem is linked with a given cause.

(b) Many diagnoses are the result of more than one problem

Many diagnoses are made because multiple symptoms or problems are recognized that point to a certain cause. This is readily seen in the matrix. The more symptoms or problems which point to a cause that can be identified, the greater the probability that the cause is correctly identified.

(c) Some problems are more easily identified

Some symptoms or problems may be more easily recognized than others and also may be more easily recognized when a particular cause is actually present. These can readily be identified from the matrix and the probability of identifying the problem can be calculated for either all causes or a single cause.

(d) Operator identification may or may not be correct

The operator may either correctly identify, incorrectly identify, or omit certain symptoms or problems. The probability that a symptom indicates a known cause can be computed from the matrix. The complement of this is the probability that a known cause exists, but the symptom was not recognized. This is the error of omission. The identification is in error when the problem is not indicative of the determined cause.

(e) A cause may exist that has not been previously defined

There will be cases where a failure occurs which has not previously occurred, thus it will not be included in the list of expected causes. No expert system or method of dealing with uncertainty can respond to this situation any more than a human expert can anticipate situations occurring that have not previously occurred, or for which no evidence exists that indicates the potential for their occurrence. The probability that the problem or symptom pointed to a cause other than the previously recognized cause could be computed, but this value would mean very little. This is true, since when the knowledge base is updated, the new cause is added. When the new cause is added to the knowledge base, the probability of finding another previously unknown cause is reduced. The important thing is that a diagnostic expert system should be capable of 'learning' from this occurrence and that it will proceed with making the best assessment of the cause in spite of this situation.

In reality all that human experts or expert systems can do in this instance, is to 'learn' that this cause exists, and update the knowledge base to account for it in the future. Using the database as a source of knowledge, the cause and problems or symptoms which were observed will become part of the expert systems knowledge. By updating the matrix regularly, the number of unknown causes can be minimized.

(f) A symptom or problem may appear that has not been observed

As with a cause, it is possible that an operator may identify a new problem or symptom. This symptom may point to an existing or a new cause. As in the case of a new cause, this information expands the knowledge base for future diagnoses, but represents only an unknown or degree of uncertainty to the current decision. Also, as in the case of the previously unknown cause, a probability of a previously unknown problem or symptom occurring could be computed, but this value would mean very little. The new problem or symptom could point to a previously recognized cause or a new cause. When the knowledge base is updated, the new problem or symptom is added, and again the probability of finding another previously unknown problem or symptom is reduced.

(g) Multiple causes may have similar certainty

More than one cause may be present that cannot be readily distinguished. The matrix solution will rank all of the possible causes of any problem or problems by computing certainty levels. If more than one cause exhibits an essentially equal, but relatively high level of certainty, it can be dealt with in one of the following ways?

1. The matrix may suggest some other problems or symptoms that were not identified. The operator could be queried about these symptoms to see if the knowledge can be increased which will make the decision process more definitive.
2. The repair person could be dispatched with the necessary tools and materials to deal with a greatly limited number of likely causes.
3. The repair person may have knowledge or experience to enhance the probability of correcting the initial diagnosis.

(h) Sufficient information may not be present for identification

The information given may not be sufficient to permit a clear choice of the cause. This would be represented by all of the causes indicating a low and relatively uniform level of certainty. If this situation occurs, the expert system has failed to provide any definite assistance in defining the problem. This would be true of any expert system regardless of the method used to deal with uncertainty.

Again, as above, the matrix may suggest some other problems or symptoms that were not identified. The operator could be queried about these symptoms to see if the knowledge can be increased which would make the decision process more definitive. Failing gaining new information, the repair person may have knowledge or experience to enhance the probability of correcting the initial diagnosis.

Generating information from the data

A number of 'relative weights' can be inferred from the matrix. These values represent a kind of 'level of understanding' of the problem. Thus, these relationships are heuristic in nature rather than mathematical. Although these values are computed using the same formulae as Bayesian probabilities, they are used as indicators of level of confidence rather than true probabilities. These relative weights are discussed below.

The relative weight that C_j is the cause

The relative weight (rw) that C_j is the cause can be computed for the known causes. It is equal to the number of occurrences of C_j divided by the total number of observed failures C_T.

$$rw(C_j) = \frac{N(C_j)}{N(C_T)} \tag{6.15}$$

For example, $rw(C_1) = 30/100 = 0.30$. This value indicates that if nothing else was known, except that the answer most likely lies within the matrix, the relative weight that the cause is C_1 equals 0.30 on a scale of 0 to 1.

The relative weight that P_i is identified as a problem or symptom

The relative weight that $[P_i|(P_i|C_j)>0]$ is identified as a problem or symptom of the cause C_j, where P_i has been identified as a symptom of C_i. This is equal to the number of occurrences of correct P_i divided by the total number of correctly identified causes C_j for which P_i is a symptom.

$$rw[P_i|(P_i|C_j) > 0] = \frac{N(P_i|P\ correct)}{\sum_{j=1}^{M} N(C_j|(P_i|C_j) > 0)} \tag{6.16}$$

For example,,

$$rw[P_1|(P_1|C_{j,j=1,\ldots,4}) > 0] = \frac{40}{20+40} = 0.67 \qquad (6.17)$$

The value computed indicates the level of confidence that problem P_i will be identified by the operator when it is known to be indicative of the cause. Table 6.6 shows the data used for the computation.

Table 6.6 Sample computation of relative weight of P_1

	P_1 $N(P_1)$=**40**	P_2 $N(P_2)$=60	P_3 $N(P_3)$=70	
C_1	$N(P_1, C_1) = 0$	$N(P_2, C_1) = 16$	$N(P_3, C_1) = 28$	$N(C_1) = 30$
C_2	$N(P_1, C_2) = 12 > 0$	$N(P_3, C_2) = 0$	$N(P_3, C_2) = 15$	$N(C_2) =$ **20**
C_3	$N(P_1, C_3) = 28 > 0$	$N(P_2, C_3) = 32$	$N(P_3, C_3) = 25$	$N(C_3) =$ **40**
C_4	$N(P_1, C_4) = 0$	$N(P_2, C_4) = 10$	$N(P_3, C_4) = 0$	$N(C_4) = 10$
Incorrect C	$N(P_1$, not C$) = 0$	$N(P_2$, not C$) = 2$	$N(P_3$, not C$) = 2$	

Values in bold type relate to equation 6.17.

The relative weight of identifying P_i given that C_j is the cause
The relative weight of identifying P_i given that C_j is the cause is equal to the number of occurrences of P_iC_j divided by the number of occurrences of C_j:

$$rw(P_i|C_j) = \frac{N(P_i, C_j)}{N(C_j)} \qquad (6.18)$$

For example, $rw(P_2|C_1) = 16/30 = 0.53$. This function can be called the 'measurability of the problem' because it indicates the level of the ability to recognize P_i as the problem when C_j is the cause. There is a maximum of two rules for each non-zero cell in the matrix.

Since more than one problem can simultaneously and independently be identified as pointing to a cause, the sum of the relative weights P_i of C_j is not necessarily equal to 1. This is why the conventional probability approach may not be applicable. Recall that conventional probability would require the relative weights to add to 1.

The relative weight that C_j is the cause given that P_i is identified as the problem

The relative weight that C_j is the cause given that P_i is identified as the problem or symptom is equal to the number of occurrences of P_iC_j divided by the number of occurrences of C_j:

$$rw(C_j|P_i\ correct) = \frac{N(P_i, C_j)}{\sum_{j=1}^{N}(P_i, C_j)} \tag{6.19}$$

For example, rw $(C_1|P_2$ correct$) = 16/(16 + 0 + 32 + 10) = 0.28$. This calculation represents the 'likelihood of the cause'. It indicates how likely the existence of a given cause is when P_2 is known to be the problem or symptom.

6.8 DATABASE REQUIREMENTS FOR THE MATRIX

Basu and Ahad (1992) discuss the use of a relational database in knowledge-based systems. They noted the large computational and I/O overhead that can result from the management of large databases that are linked to knowledge-based systems (Hu, 1989; Marquess, 1991; Moore, Schussler and Shafer, 1989). The premise of the proposed methodology is that such costly overheads can be reduced by using expert heuristics to prune the problem–cause relationship data space. We present the general requirements of databases which will provide the necessary information to support the heuristic matrix approach.

6.8.1 Database systems for maintenance operations

It is anticipated that the major application of the heuristic matrix expert system will be improvement of equipment maintenance in a manufacturing environment. Maintenance operations in manufacturing are usually quite spread out. Equipment outages are generally reported by an operator or operating supervisor from a remote location to a central dispatcher or control room facility and communication is by radio, pager phone system, fax or electronic mail. The dispatcher normally logs the outage. This may be done by a handwritten report or entered into a computer. At a minimum, this record will include the equipment number of the affected machine, the time of the outage and some delay code or operator description of the malfunction. The dispatcher then assigns the appropriate repair personnel or enters the job into the queue if none is available.

Systems of this type have been in use in maintenance operations for many years (Parsaye *et al.*, 1990; Waterbury, 1990). Until computers were introduced into the system, the only records were entries in log books which were usually just notes written in the dispatcher's words. Often, these records were only tracked until the repair was completed. Since computers were introduced into the system in the late 1950s and early 1960s, the desire for longer term record keeping and analysis of the information has increased. Many companies developed systems for logging outages. This was initially accomplished by notes taken on forms, then later entered into the computer by a keypunch operator.

It is now common for operations to have control rooms where control room operators enter the outage reports directly into the computer. In many cases, work orders are generated automatically. When the work order is created, a job number is assigned, and this number is used to schedule work, track backlogs and track all labor and material used to make the repair. Recently, systems have been introduced by which the equipment operator can enter an outage directly into the computer. Two major categories of delays are reported to the control room.

- Operational or production delays: the operation is halted by operational problems, not of a maintenance nature, at least not due to maintenance on the equipment or system being delayed.
- Maintenance delays or outages: the machine is out of service during operations due to a mechanical or electrical malfunction or failure of the equipment or system.

Normally, when a delay is reported to the control room or entered into the system it will be designated as either production or operational in nature. In many systems, if the delay is operational, the data is entered into an operational database for future analysis. If the delay is of a maintenance type, the information is either entered into the maintenance work control system or more likely, a work request is generated, and this work request is entered into the maintenance work control system. Most often, this is done in lieu of an entry into the operating system.

In the case of a system which the first author helped to install, these entries were made so that future analysis of the impacts of the outage on other portions of the operation could be made. This portion of the system was later discarded since no use was being made of the information. Operations management was concerned that often equipment operators would not just report the problem, but would enter their diagnoses of the cause. This very often differed from what the maintenance personnel determined to be the cause and the conflicting reports were considered to be an embarrassment to management.

6.8.2 Minimum information requirements

The data management system necessary to support the heuristic matrix expert system requires that both operating and maintenance records be kept on all outages and breakdowns. The capability must be inherent in the databases to correlate what the operator reported was wrong and what maintenance was performed. Often the information reported by the operator is not retained in the case of maintenance outages. To make it possible for the data to be used to produce rules, the machine identification and the operator's assessment of the problems encountered must be logged in the operator's report of outages in the management information system. It is not necessary that the operator's diagnosis of the cause be entered or retained. The purpose of the diagnostic expert system is to bridge the gap between the description of problem and the determination of the cause, based on historical knowledge of what was done in the past to remedy the situation. To be usable for creating rules for the matrix system, it is, however, necessary that a method be available so that this information can be correlated with the work control system. The best way to accomplish this would be to include the work request number in the operating system database entry.

Maintenance activities are identified, planned and scheduled, and the work performed is recorded. The work, which can be tracked on the system, would generally be expected to include all predictive, preventive and corrective maintenance performed on all the major equipment and systems in the operation. In practice, the degree to which the systems are utilized and the level of detail entered into the system varies from operation to operation.

The two integral portions of the maintenance system upon which the ability to generate rules for the matrix system is dependent are the work request system and the equipment history databases. The work request system is necessary because the work request number provides the link between the operations report and the maintenance performed as discussed previously. The equipment history databases are the source of the knowledge of what remedial action was actually taken. The level of detail contained in these databases will vary. The minimum information necessary for the matrix rules discussed above is a basic description of what component or system was affected. For example, 'the traming motor main bearing was replaced'. A later section discusses additional research for expansion of the system which could be developed if more detailed information was recorded in the equipment history databases.

6.8.3 Suggested database structures

It is not the intent of this chapter to suggest total removal of human intervention from the process. Instead, the idea is to make use of the large volume of information which operators and managers record about their operations. One maintenance operation averages over 20 000 entries each year into the operating and maintenance databases describing operational and maintenance delays and outages. This data includes much more information than the minimums described above, since not all of the capability which was designed into the system is implemented. The process of generating rules will begin with merging and sorting the information to determine the input values for the matrix presented earlier. It is anticipated that proper determination of which problem or cause category some of the entries should be placed will require some information from the operators or maintenance personnel involved. Using the databases described in this section, these pieces of information can be easily identified. Other possible uses for the additional data are discussed later. The following paragraphs will describe the recommended type of databases and their content. The databases described will be similar to those which were designed for the system at the large integrated maintenance operation mentioned previously. The operations MIS system contains the following types of databases:

1. Personnel databases
   (a) personnel data
   (b) training records
2. cost accounting system
3. warehousing and inventory control
4. operations control and tracking
   (a) system databases
   (b) production/operation
   (c) injury report
5. integrated maintenance control system
   (a) work request and scheduling system
   (b) maintenance equipment history databases.

In all of the above databases are 'key fields' associated with the others through which all personnel, materials used, action taken and cost can be linked together. Entries into these databases are made by appropriate personnel usually via pop-up forms which query the person for the required information. Where possible, default value cross-referencing and other data input validation techniques are used to ensure that the data inputs are correct.

6.8.4 Operations tracking and control

The databases, from which data must be retrieved for use in generating rules, are primarily those in the operations and control group. Some crossover with warehousing and inventory control will be desirable to help determine or fix the cause. A chart showing the suggested structure of the operations control and tracking system is shown in Fig. 6.2. The system databases are presented below.

- Equipment lists – these contain equipment number, vendor, type, model, etc.

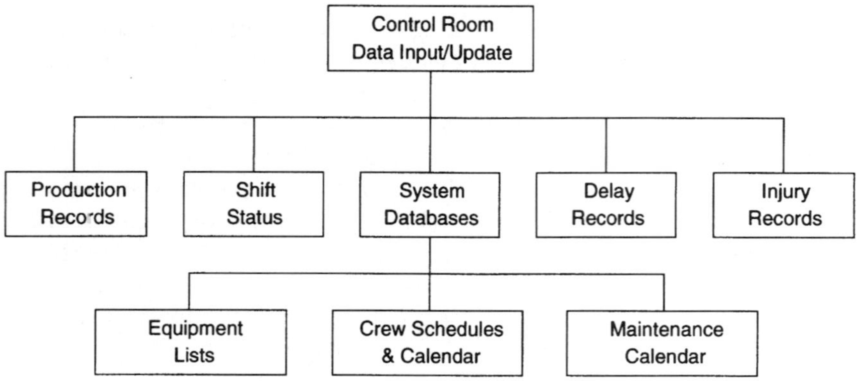

Fig. 6.2 Operations tracking and control databases.

- Crew schedules and calendar – this will present the personnel and equipment they are operating or their job assignment.
- Maintenance calendar – this is generated by the maintenance system and shows what equipment is scheduled down for preventive maintenance. It is also used to flag changes in crew schedules and calendar and equipment utilization schedules.
- Crew lists – these show personnel assigned to an operating crew.
- Personnel calendar – this contains normal work schedules for personnel; its purpose is to flag exceptions in crew schedules and calendar.
- Numerous small databases, which will provide the choices, will be presented to a user on pop-up menus used to aid accurate data inputs.

At the beginning of each shift, a record is generated in the crew assignment list database in the control room. The dispatcher or operating supervisor enters the initial shift status. The crew assignment list database is the main database of this group. Information is drawn from the other databases in the group and is used to provide detailed correlation between the operation/production databases, the specific personnel and equipment involved.

Most of the information is entered by default, when the crew identification code is entered. Changes need only be made for exceptions to the calendars and crew. This designation will not have to be altered during the shift unless a crew, equipment or job assignment is made. The crew identification code is the primary key which will correlate this data with the other production and maintenance databases. A secondary key field entry number will be generated by the system. It will be zero for the initial entry and incremented if additional records are entered with the same crew identification code during the same shift.

6.8.5 Production and operations databases

This group consists of a production log database and a delay log database. The production log database contains the date and shift, crew identification code, entry number and all the various units of work accomplished. The current crew identification code and entry number would default to the record to ensure proper correlation with personnel and equipment later. The delay log, which is the one that is of interest for obtaining the matrix data, will contain all the information related to a given delay or outage.

6.8.6 Generating a maintenance delay record

When a delay occurs, a record will be initiated by either the control room, dispatcher or operator as described earlier. When the delay entry area is entered, a code identifying the nature of the delay either production/ operations or maintenance will be entered, and the date and time will default from the system clock. The system will begin querying using pop-up menus to further define the delay. The process will be similar whether the delay is production/operations or maintenance. Since maintenance delays are the ones which are of initial concern in the development of the heuristic matrix expert system, only the generation of this type of record will be described in detail.

If the delay is a maintenance delay, a menu-driven system will query for the crew identification code. The current entry number will default into the system, and a work request number will be assigned. A pop-up menu will request the equipment assigned to the crew and the person entering the delay will select the appropriate unit. Various menus will begin popping up with lists of possible problems which have been associated with this type of equipment. Each successive menu will further define the delay, reducing it from system to subsystem to the depth which the operator can respond from observation. Each of the successive definitions adds another field to the delay identification code.

It should be noted that more than one problem may be identified. If a problem cannot be described by the menus, the delay would be labeled unknown. A field would be included in the delay record to enter a brief description of the problem. This field can also be used for details that are in addition to the delay code. An example of how these menus would build the delay code is shown in Fig. 6.3. A delay record will remain open until the cause of the problem is determined and the equipment is back on line. At this time the completion time will be entered. A sample of what the database might look like is shown in Table 6.7.

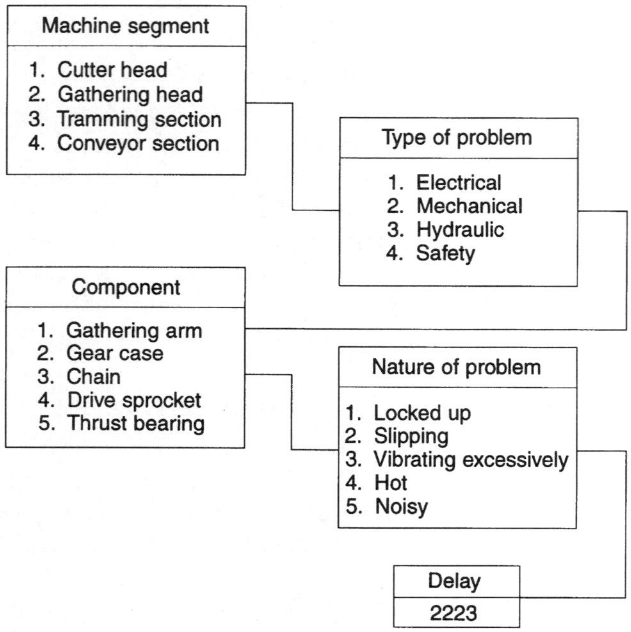

Fig. 6.3 Delay code development.

6.8.7 Determining the cause of the problem

Although this topic is not part of the definition of the databases, it should be noted that the problem or problems have now been defined. It is at this point, that the diagnostic expert system would be used to help define the cause of the problem or problems before maintenance personnel are dispatched.

6.8.8 Definitions of the maintenance databases

The maintenance databases which will provide the information for the matrix will be expected to be incorporated in any large integrated maintenance control system. This system will at least include the following major elements:

- a work request or work order system;
- a backlog control system and planner/scheduler;
- a preventive and predictive maintenance scheduler;
- a comprehensive equipment history record keeping system.

Table 6.7 Delay database sample

Crew	Type	Equip. no.	WR No.	Down Date	Down Time	Up Date	Up Time	Delay code	Comments
5A1	M	CM01	1239	03/01/93	17.23	03/01/93	17:27	45546456	
5C3	M	CM03	3356	03/21/93	03:09	03/21/93	03:15	566576	
5D1	M	CM04	4542	03/22/93	04:11	03/22/93	04:25	7657658	
5C2	M	CM03	7678	03/25/93	17:25	03/25/93	18:09	243434	
5D1	M	CM03	5652	04/02/93	19:20	04/02/93	19:30	6565476	
5C1	P	CM02	6567	04/03/93	07:24	04/03/93	09:30	34234	
5D1	P	CM03	8787	04/06/93	17:02	04/06/93	17:55	565876	
5B2	P	CM01	9894	04/10/93	15:23	04/10/93	16:25	4345	
5A2	M	CM03	1334	04/15/93	17:03	04/15/93	18:00	7878	
5C1	M	CM01	3435	04/16/93	17:23	04/16/93	17:29	45454	
5B1	M	CM03	6565	04/16/93	17:30	04/16/93	20:30	12111	
5A3	M	CM05	7678	04/16/93	21:40	04/16/93	22:45	1122	
5A3	M	CM03	7897	04/17/93	08:37	04/17/93	09:30	656567	
5D1	M	CM01	9442	04/25/93	19:21	04/25/93	20:22	2323	
5A1	M	CM01	1112	05/05/93	20:25	05/05/93	20:45	6576879	
5D1	P	CM03	2334	05/20/93	07:22	05/20/93	09:25	43434	

A schematic diagram of a system of this type is shown in Fig. 6.4. The work request and equipment history portions of the system are the areas of interest in defining the cause associated with the problems reported into the delay log database.

6.8.9 Determining the cause of a delay

The work request numbers which were generated by the delay reporting system are the key to correlating the operators' reports with the equipment history databases to find the cause of the reported delay. Two of the sources of work requests shown in Fig. 6.4 will include the operator assessment of the problem, breakdown reports and end of shift operators' reports. The definition of the cause will come from the stated cause, component number and list of parts taken from inventory. Examples of fields in the equipment history database are described below.

- EQID refers to the equipment identification type and number, CM stands for continuous miner.
- INIT DATE refers to the date the work request was initiated.
- COMP DATE refers to the date the work request was completed.
- FMAN is the initials of the reporting supervisor.

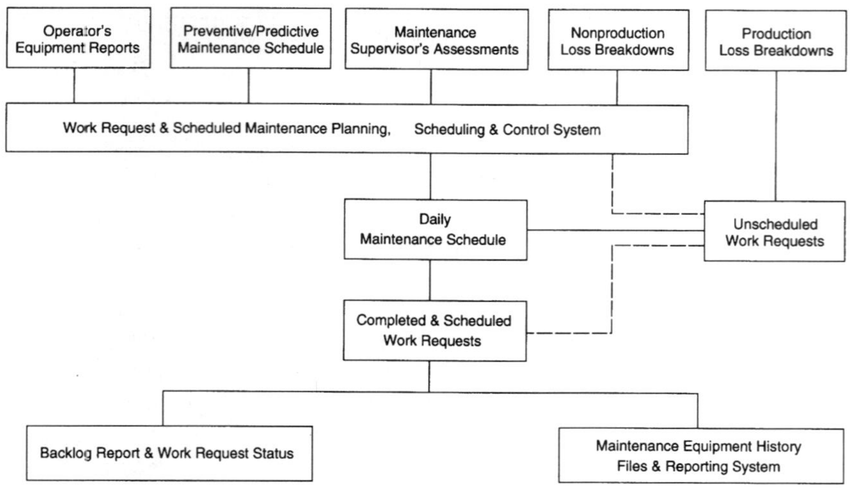

Fig. 6.4 Schematic of a maintenance control system.

- WKRREQ is the work request number.
- AHRS is the actual time to repair.
- COMP refers to an identification number given to major components.
- WORK REQUEST/COMPLETE SUMMARY is a brief description of the work preformed.

6.8.10 Cause-of-delay database

The data in the equipment history databases will probably require interpretation before the cause can be defined in a usable manner. This equipment history file defines the cause to the level of the component number, the cause itself is given in the mechanic's description of this work. This literal description of the cause will have to be transformed into an additional digit in the component code. This can be done by reviewing the list of causes and adding the action digit by interpreting the mechanic's description and assigning the appropriate number. This is accomplished in a manner similar to the way problems are defined in the delay log database. A person familiar with the equipment such as the maintenance engineer or planner would then review the information. A pull down menu, the type used to define the problem or problems (Fig. 6.3), would then be used to develop a cause code. It is likely that an additional database would need to be developed. This database would include the work request number, some of the other information from the equipment history and the four-digit cause code. The database can be developed using the same database manager which manages the system and production/operations databases.

6.8.11 Establishing new problem or cause codes

Periodically, operations and maintenance personnel must review the problems or causes coded as unknown in their respective databases. They will have to be grouped manually and new codes assigned to these previously undefined groups. Two databases with a common key field are required for the system to work:

- the delay log containing the maintenance delays;
- the delay cause database.

The databases described previously would be the best media from which to extract the parameters needed for the matrix. If, however, the minimum fields of data described above are available in two databases, the system can be made to work. The minimum databases could be developed and entered manually.

6.8.12 Frequency of identification of problem P_i

To reduce the databases to find the data required for input into the matrix, the following steps would be taken.

- Sort the delay log and then retain only maintenance delays.
- Sort the remaining portion by equipment type and retain the record dealing with the equipment for which the matrix is being constructed.
- Count the number of each delay code $N(P_i)$.

Many state-of-the-art database managers will permit this operation to be accomplished in a single step, a query to produce the count of the code field by code for the selected type of equipment.

6.8.13 Frequency of identification of cause C_j

The values for $N(C_j)$ can be determined by performing the same process as described for determining $N(P_i)$ on the delay cause database.

6.8.14 Frequency of when P_i was indicated as the problem when C_j is the cause

This frequency is denoted as $N(P_iC_j)$. These values can be determined by crossing the two databases and first sorting the resulting database by the cause code and then counting the number of occurrences of each problem P_i when the cause C_j is present.

6.9 GENERATING THE KNOWLEDGE BASE

In this section, a model containing a simple sample heuristic matrix and a knowledge base is presented. The data in the sample database is not complete because some of the information was not entered or was not saved. Thus, using the available data will produce a somewhat simplistic example because only the maintenance personnel's after-the-fact interpretation of the problem/symptom is generally stated as the exact explanation of the cause. For example, the problem/symptom is the 'left cutter motor is bad', and the cause/action is 'replace cutter motor'. This data, however, can demonstrate the problem–cause relationship and provides frequency of occurrence of failures of some major component types.

6.9.1 Expert system development tool

VP-Expert, a commercial expert system shell, was used as the development tool for the sample development. This shell has the ability to produce a knowledge base from data stored in external databases or spreadsheets. The software also allows the user to include certainty factors into the rules produced. Although providing an adequate avenue for this demonstration model, it has a number of shortcomings which would make it, and many other commercial expert system development tools inadequate for the purpose of the heuristic matrix diagnostic expert system. These shortcomings include the following:

- The induction table does not permit the automatic inclusion of certainty parameters in the rules.
- It does not allow the creation of rules with multiple conclusions with different certainty:

$$\text{IF } P_i \text{ THEN } C_1 \, [\text{cv}_1], C_2 \, [\text{cv}_2] ,..., C_N \, [\text{cv}_N]$$

- It uses only the MYCIN-type certainty factor approach in handling uncertainty. Thus, although the model demonstrates the method, the certainty values are distorted.

6.9.2 Database manipulation and matrix construction

Because the database is relatively small, it is possible to do the necessary sorting and frequency tabulation in Lotus 1–2–3. Using this software permits the database manipulation, the heuristic matrix and the VP-Expert induction table to be constructed in a single spreadsheet. An actual system would make a more comprehensive database manager. Many adequate database management systems are commercially available. The approximately 200 data samples were sorted and summarized into a refined database. This provides the input data for the heuristic matrix.

6.9.3 The heuristic matrix

Using the problem-cause data and frequency distribution, the heuristic matrix shown in Fig. 6.5 is created. Notice that the numbers within the cells of the matrix add up to the frequency values in the columns and rows for problems and causes, respectively. For example, 'foot shaft' was identified as the problem 30 times. Eight of those times, the repair action was 'replace foot shaft' while 'repair foot shaft' was performed 22 times. Using the matrix, the relative weights can be calculated using the equations presented by Peters (1992). These relative weight values are then used to compute the likelihood function [Lcv] and the measurability functions [Mcv] can be computed using the equations procedure presented by Peters (1992). The certainty value for likelihood rules is

	PROBLEM IDENTIFIED													
	Cutter Motor	Cutter Clutch	Tram Motor	Conveyor Motor	Pump Motor	Sheer Jacks	Planetaries	Conveyor Chain	Flexboard	Foot Shaft	Ripper Chain	Drum	Bit Blocks	
Freq	14	15	15	11	7	11	21	15	12	30	7	2	20	CAUSE OR REPAIR ACTION
5	5													Replace Cutter Motor
9	9													Repair Cutter Motor
8		8												Replace Cutter Motor Clutch
7		7												Repair Cutter Motor Clutch
3			3											Replace Tram Motor
12			12											Repair Tram Motor
8				8										Replace Conveyor Motor
3				3										Repair Conveyor Motor
4					4									Replace Hydraulic Pump Motor
3					3									Repair Hydraulic Pump Motor
6						6								Replace Sheer Jacks
5						5								Repair Sheer Jacks
8							8							Replace Planetaries
13							13							Repair Planetaries
3								3						Replace Conveyor Chain
12								12						Repair Conveyor Chain
2									2					Replace Flexboard
10									10					Repair Flexboard
8										8				Replace Foot Shaft
22										22				Repair Foot Shaft
3											3			Replace Ripper Chain
4											4			Repair Ripper Chain
2												2		Replace Drum
20													20	Replace Bit Blocks

Fig. 6.5 Heuristic diagnostic matrix.

denoted as Lcv (likelihood confidence value). The certainty value for measurability rules is denoted as Mcv (measurability confidence value). This indicates the discriminating capability of the matrix in identifying problem–cause relationships.

6.9.4 Rule induction

The VP-Expert induce command creates rules from columns of data. The left-most column represents the premise of the rule and the right-most column represents the conclusion. Additional columns between them are used to qualify the premise. These qualifiers follow the premise and are connected by the logical AND. Since VP-Expert does not provide for multiple conclusions, two rules will have to be induced for each filled intersection in the matrix, a likelihood rule and a measurability rule. The induction table for the model will have only two columns, the problem/symptom and the cause/repair.

6.10 CONCLUSIONS AND RECOMMENDATIONS

Interest in the potential benefit of expert systems is increasing in the manufacturing. This chapter demonstrates the viability of the matrix diagnostic expert system. It is anticipated that this, coupled with the increased level of industry acceptance, will provide incentive for further research. Some potential directions for further research with the approach presented are discussed below.

6.10.1 Definition of the functions

The exact nature and shape of these functions would have to be determined from a very large database, much larger than was available for the sample problem. The resulting equations could be complex or quite simple, and the relationships used in the test model may function correctly, or on the other hand, a more extensive database might cause them to be revised for this application. This may be a revision in structure of the equation or the addition of constants which would shape the function. If shaping constants are required, they may vary with other types of equipment.

6.10.2 Establishing criteria for minimum acceptable measurability

The measurability value is a measure of the relative ability of an operator to recognize a problem P_i which is indicative of cause C_j. If this value is low, the acceptance of problem P_i becomes suspect. Liang (1992) discusses the use of hypothesis testing, applying 'hurdle values' to determine whether sufficient evidence is available to make the suggested reference. A test could be defined which would determine whether a connection between the problem and cause has a measurability value high enough to provide an acceptable level of confidence.

6.10.3 Rule interaction

It has been previously stated that 'measurability rules' are measures of the ability to recognize P_i, as a problem associated with C_j. In the model, the relationship between the measurability and the likelihood is assumed to be a simple multiplication. In this case, the 'measurability rules' will perform a very direct modifying function to the 'likelihood rules'. It is assumed that the 'measurability rules' are somewhat subservient to the 'likelihood rules'. The exact way in which these functions interact with one another, like the measurability and likelihood functions themselves, would require a substantial database.

6.10.4 Low level of identification confidence

In some cases, the confidence value associated with the resulting 'most likely' conclusion is too low. A test would need to be designed which would force the system to search other paths for a possible greater correlation level. This could possibly be accomplished by one matrix returning the process to another matrix in the hierarchy to attempt another path. It is possible that all logical paths can be followed to the end where the confidence can be tested and the 'most likely' or multiple logical alternatives can be suggested.

6.10.5 Additional research on applications

The potential enhancements which are discussed in this section would most likely be dependent on a higher level of information being recorded and maintained in the equipment history databases. They are, however, plausible ways in which the value and utilization potential could be increased greatly. Most of them would require a substantial effort to accomplish. The additional requirements would include such things as adding fields to the equipment history which record the levels of variables associated with the problem. In other cases, the additions might be in the form of links to other databases such as personnel records, cost system data or warehouse issues records.

6.10.6 Recommending parts, tools and equipment

Once a 'most likely' cause or alternative causes have been predicted, additional matrices could suggest parts, special tools such as extra large wrenches, or equipment such as a boom truck that should be available to make the repair. Often, the parts are expensive and once removed from the warehouse in a maintenance environment are subject to a high risk of loss if not actually used in the repair. Thus, this concept might be further expanded to include some sort of risk analysis on very expensive parts. This would weigh the cost benefits associated with having to wait for a part to be dispatched if needed against the risk of taking the part to the repair site and not using it.

6.10.7 Checking repair personnel qualification

Often the experience, training and skill level of the maintenance personnel vary greatly. When possible, it is beneficial to match the skills of maintenance personnel with either the difficulty or the criticality of the job. This concept would require both an evaluation of the priority associated with the problem and the skills of the available personnel. Priority would require evaluation of both the requirements of safety of personnel, of the operation and the loss cost.

6.10.8 Describing simple checks and repair procedures

This system could be used to prompt an operator to take some rudimentary actions which would either correct or increase the knowledge of the problem. This could reduce the number of calls requiring dispatching of maintenance personnel, or at least improve their ability to come to the repair site with the correct parts and tools.

6.10.9 Knowledge expansion

The system could be expanded to provide lists of checks that the operator could be instructed to make by the dispatcher, which could serve to confirm the expert system's initial diagnoses of the cause of the problem. The suggested checks would be the ones which are specifically related to the problem or problems observed. The results of this additional information would either increase or decrease the confidence in the diagnoses. If the confidence is substantially reduced, the system would back up and re-evaluate the diagnoses.

6.10.10 Training for making simple repairs

Once a 'most likely' cause is determined, the system could instruct the operator to make a simple repair or adjustment. This could remedy the condition, thus eliminating the need to dispatch maintenance personnel. On the other hand, if the suggested repair fails to remedy the condition, this information is entered into the system. In this context the attempted repair becomes another element of knowledge, and is used in the same manner as the test results described above.

6.10.11 Linking matrices to expand the capabilities

One of the strengths of this method is that matrices can be linked together. They can reduce the problem from the system level to the subsystem, then to the component or part. Additional matrices can be used after a cause is determined to suggest parts, tools or equipment the service personnel may need in making the repair, or to select the most qualified available person to answer the call. This linking relationship is shown in Fig. 6.6. The rules in the system matrix determine the most likely subsystem. The inference engine can then pass the processing to the appropriate subsystem matrix, and so on.

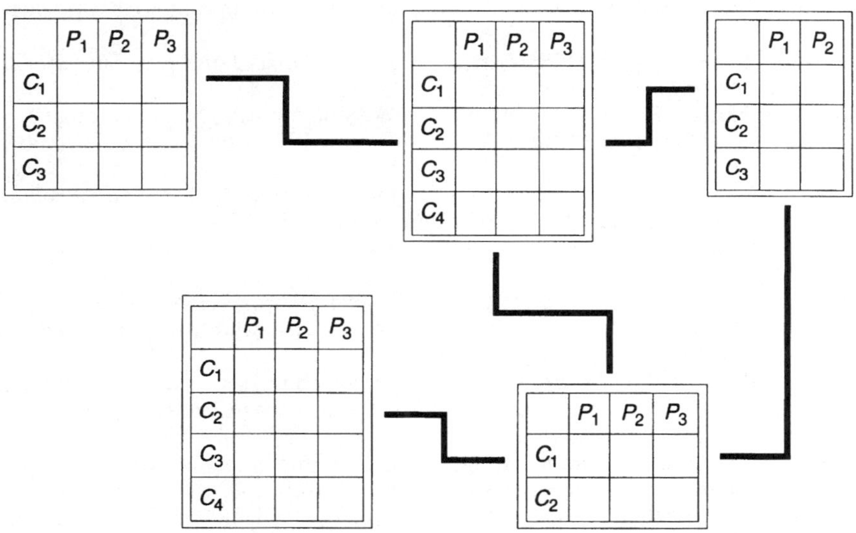

Fig. 6.6 Linked matrices.

REFERENCES

Adlassnig, K.P. (1986) Fuzzy set theory in diagnosis. *IEEE Transactions On Man and Cybernetics*, **SMC-16** (2), 260–64.

Arend, M. (1989) AI: expert keeps up with bank's currency events, *Wall Street Computer Review*, **6** (9) 22–24, 91–92.

Badiru, A.B. (1992) *Expert Systems Applications in Engineering and Manufacturing*, Prentice-Hall, New Jersey.

Basu, A. and Ahad, R. Using a relational database to support explanation in a knowledge-based system. *IEEE Transactions on Knowledge and Data Engineering*, **4** (6), 572–81.

Blackman, M.J. (1990) Case for expert systems. *AI Expert*, February, 26–32.

Boult, R (1988) ISDN: Snapshot of the French scene. *Computerworld*, **22** (50), 80.

Bowerman, R.G. and Glover, D.E. (1988) *Putting Expert Systems Into Practice*, Van Nostrand Reinhold.

Buchanan, B.C. and Shortliff, E.H. (1984) *Rule Based Expert Systems: The MYCIN Experiment of The Stanford Hueristic Programming Project*, Addison-Wesley Publishing Company, Reading, Massachusetts.

Codd, E.F. (1990) Mastering the art of database fusion. *Datamation*, October 15, 61–64.

Evans, S. (1988) A clinical tool for nursing. *Computers in Healthcare*, **9** (8), 41–44.

Falkenhainer, B. (1986) Towards a general purpose belief maintenance system, in *Proceedings of the Second Workshop on Uncertainty and Probability in Artificial Intelligence*, Philadelphia, PA.

Gorman, H. (1987) Experts are coming to put down uptime cost uprisings. *Production Engineering*, **34**(8), 68–71.

Harmon, P. and King, D. (1985) Expert systems, in *Artificial Intelligence in Business*, John Wiley & Sons, New York.

Hashemi-Nassab, R. (1989) An automated rule generator (R2) based on rough sets, in *Proceedings of the Oklahoma Symposium on Artificial Intelligence*, November, 107–20.

Hau, H.-Y. and Kashyap, R.L. (1989) On the robustness Of Dempster's rule of combination, in *Tools for Artificial Intellegence*, IEEE, 578–82.

Herrod, R.A. (1988) AI: promises start to pay off, *Manufacturing Engineering*, **100**(3), 98–103.

Hu, D. (1989) *C/C++ For Expert Systems*, MIS Press, Portland, Oregon.

Kaplan R.S. (1984) Yesterday's accounting undermines production. *Harvard Business Review*, July–August, 95–101.

Kerklof, P.L.M. (1989) Matrix structure representation of a medical knowledge base, in *IEEE Engineering in Medicine and Biology Society 11th Annual Intenational Conference*, 1832–1833.

Liang, T.-P. (1992) A composite approach to inducing knowledge for expert systems design. *Management Science*, **38**(1), 1–16.

Mahler, E.G. (1989) Keynote address, presented at Oklahoma Symposium on Artificial Intelligence, November.

Marquess, P. (1991) Object-oriented paradigm. *DEC Professional*, March, 50–59.

Merino, D.N. (1989) Developing economic and noneconomic models incentives to select among technical alternatives. *The Engineering Economist*, **34**(4), 275–90.

Mitchell, D.H. (1988) The Dempster–Shafer theory of evidence, a model of human decision making, in *Proceedings of the Oklahoma Symposium on Artificial Intelligence*, November, pp. 235–40.

Moore, A., Schussler, T. and Shafer, D. (1989) *VP-Expert, Rule-Based Expert System Development Tool*, Paperback Software International.

Morawski, P. (1989a) Bayesian belief networks explained. *AI Expert*, May, 44–48.

Morawski, P. (1989b) Programming Bayesian belief networks. *AI Expert*, August, 74–79.

Myers, R.C., Jr (1988) AI and laser disk improve maintenance workstations, *Automation*, 35(3), 40–41.

Parsaye, K., Chignell, M., Khoshafian, S. and Wong, H. (1990) Intelligent databases, *AI Expert*, March, 38–47.

Pearl, J. (1988) *Probabilistic Reasoning in Intelligent Systems: Networks of Plausible Inference*, Morgan Kaufmann, San Mateo, California.

Peters, J.W. (1992) *Generating knowledge base information from MIS databases for equipment diagnostic expert systems.* PhD dissertation, School of Industrial Engineering, University of Oklahoma, Norman, Oklahoma.

Raiffa, H. (1968) *Decision Analysis: Introductory Lectures on Choices Under Uncertainty*, Random House, New York.

Rowan, R. (1989) Diagnostic systems for manufacturing, *AI Expert*, April, 28–39.

Schank, R.C. and Abelson, R.P. (1977) *Scripts, Plans, Goals, and Understanding, Erlbaum*, Hillsdale, NJ.

Sumita, U., Masuda, Y. Tachi, M. and Ishikawa, Y. (1989) Structural analysis of the Dempster–Shafer theory and related limit theorems. *International Journal of Expert Systems*, **2**(2), 163–98.

Thompson, B. and Thompson, W. (1986) Finding rules in data. *Byte*, November, 149–60.

Waterbury, R.C. (1990) Intelligent databases raise expert system I.Q. *Intech*, December, 46–47.

Wilhelm, M.R. and Parsaei, H.R. (1991) A fuzzy linguistic approach to implementing a strategy for computer integrated manufacturing. *Fuzzy Sets and Systems*, 191–204.

Zadeh, L. (1992) The calculus of fuzzy IF/THEN rules. *AI Expert*, March, 22–27.

CHAPTER SEVEN

Object-oriented organization of network flow problem-solving knowledge for manufacturing decision-support systems

Prashant Banerjee and Qais Barakat

7.1 INTRODUCTION

A network representation is commonly found in areas such as production, distribution, project planning, facilities location, microelectronics, resource management and financial planning. The network flow problem-solving knowledge is a class of network problems which can be used to optimize certain problem parameters. As an illustration the shortest path problem, the minimum spanning tree problem and the maximum flow problem are studied in this chapter.

The chapter is subsequently divided into six sections. Some of the applications of the proposed material are given in section 7.2. The object-oriented paradigm is discussed in section 7.3. The targeted network flow problems are introduced in section 7.4. The object-oriented design issues are addressed in section 7.5. Implementation and observations are addressed in section 7.6. Finally, conclusions are presented in section 7.7.

7.2 APPLICATIONS IN MANUFACTURING DECISION SUPPORT SYSTEMS

The network flow problem-solving knowledge has a number of applications in manufacturing decision support systems, especially in the area of design leading to manufacturing specifications. The **minimum**

spanning tree is used in planning transportation networks. The network maps the locations that are required to be accessed by the transportation lanes (highways, railroad tracks, air lanes and so forth); and the costs of providing the transportation lanes. In this context, the minimum spanning tree problem is to determine which transportation network would service all the locations with minimum total cost. Other examples where the minimum spanning tree is used are in designing cable layouts, transportation networks and in wire length estimation for laying out integrated circuits prior to manufacturing. The **maximum flow** algorithm is used in designing many microelectronic devices to prepare their manufacturing specifications. The maximum flow algorithm is also used in other models such as the **cut tree**. The cut tree has been used as a valuable tool for supporting the facility designer in analysis, construction and improvement of facility layouts. Generally the cut-tree-based layout methodology has permitted the generation of satisfactory design with flow distance improvement on the order of 25–55% over the best alternative available.

If an object-oriented methodology is used to code solutions for such network flow problems as an alternative to the classical functional methodology, a substantial reprogramming effort can be avoided and consequently programmer productivity can be improved. In an object-oriented paradigm, there is a standard library of objects such as the object library in Smalltalk (Goldberg and Robson, 1983) and C++ (Lippmann, 1988) programming languages. Unfortunately such a library is not efficient in dealing with network flow representation. Hence an additional **library of objects** is created which can efficiently handle network flow problems. In this chapter the design issues for such a library of objects are analyzed.

7.3 OBJECT-ORIENTED PARADIGM

The object-oriented paradigm is based upon the following primary concepts: objects, attributes, classes and services. They are outlined below. More details can be found in the many excellent general references available on the subject (Booch, 1986; Bourne, 1992; Coad and Yourdon, 1991; Goldberg and Robson, 1983; Meyer, 1988; Wilde, Matthews and Huitt, 1993) and its applications in manufacturing (Adiga, 1989; Lenard, 1993; Montreuid and Banerjee, 1988; Muhanna, 1993.

An **object** is a programming entity which has a set of data and procedures encapsulated in it. The data or state information for each object is stored as **attributes**. In object-oriented programming an attribute is also referred to as instance variable. The procedures encapsulated in an object are referred to as **services**. An object interacts with another object through passing **messages** which are calls using the services to those of another object. A **class** is a description of one or more objects with a uniform set of attributes and services.

7.3.1 Significance of object-oriented paradigm

Some of the significant features are as follows.

1. Modeling the real world as close as possible to a user's perspective. Objects in the program match objects in the real world closely, so real world changes should be easier to map in terms of program modifications.
2. Explicit representation of commonality. The object-oriented paradigm uses inheritance to identify and capitalize on commonality of attributes and services.
3. Splitting of large systems into smaller and manageable segments. The number of decomposed units in an object-oriented decomposition is usually much higher than in a functional decomposition.
4. Construction of reusable software components and easily extensible libraries of software modules.

7.4 NETWORK FLOW PROBLEMS

7.4.1 Terminology

Network flow problems have been discussed in depth in many textbooks on operations research (e.g. Hillier and Lieberman, 1990). A network consists of a set of **nodes** (vertices) and **links** (edges, arcs). From the standpoint of network flow, nodes are of three types: source node, sink node, and transshipment node. A node is termed a supply node if the cumulative flow out of the node exceeds the cumulative flow into the node and it is termed a sink node if the cumulative flow into the node exceeds the cumulative flow out of the node. For the transshipment node the cumulative 'flow in' is equal to the cumulative 'flow out'. The edges of a network may have a flow of some type through them The flow may be directed or undirected. If the flow is allowed in only one direction, the edge is called directed; otherwise the edge is said to be an undirected edge. The maximum amount of flow that can be carried on a directed edge is referred to as the edge capacity. The network is called a directed network if all the edges are directed. Similarly if all the edges are undirected, the network is called an undirected network. A path between two nodes is a sequence of distinct edges connecting these nodes.

7.4.2 The maximum flow problem

In a maximum flow problem, the edges have a limited capacity. The objective is to determine the feasible pattern of flows through the network that maximizes the total flow from the source node to the sink node. An **augmented path** algorithm (Bazaraa, Jarvis and Sherali, 1990) is used for solving this problem. This algorithm is based on two intuitive concepts, those of a residual network and of an augmenting path. While details are available in many standard books on linear programming (e.g. Bazaraa, Jarvis and Sherali, 1990), an overview of the approach is given here to help explain the developed objects in the library.

In the solution methodology for maximum flow problems, two networks are used. One is the original network representing the problem and the other is a residual network formed from the original network. A residual network differs from the original network only in that each directed edge ($i \rightarrow j$) lacking a directed edge in the opposite direction ($j \rightarrow i$) now has such an edge added with zero edge capacity. Following this, the edge capacities in the residual network (called residual capacities) are adjusted as follows with each iteration. Every time a flow amount Δ is added to edge $i \rightarrow j$ in the original network, the residual capacity of edge $i \rightarrow j$ is decreased by Δ and the residual capacity of edge $j \rightarrow i$ is increased by Δ. Thus the residual capacity represents the unused arc capacity in the original network or the amount of flow in the opposite direction in the network that can be canceled. Therefore, after assigning various flows to the original network, the residual network shows how much more can be done either by increasing flows further or by canceling previously assigned flows.

An augmenting path is a directed path from the source node to the sink node in the residual network such that every edge on this path has a strictly positive residual capacity. The amount of flow that can feasibly be added to the entire path is called the residual capacity of the augmented path, which is the minimum of the residual capacities of the edges in the augmented path. The augmented path algorithm repeatedly selects some augmented path and adds a flow equal to its residual capacity to the path in the original network. This process continues until there are no more augmenting paths, so the flow from the supply node to the demand node cannot be increased further and this series of iterations is called a **round** in our terminology.

Next, we come to the notion of a **cut**. A cut can be defined as any set of directed edges containing at least one edge from every directed path from the source node to the sink node. The utility of a cut between the source node and a sink node in a graph is to partition of all nodes into two distinct subsets, with one subset containing the source node and the other subset containing the sink node. The value of a cut is defined as the sum of the weights associated with all edges having one node in each subset. A minimum cut between a pair of nodes is a cut separating these nodes that, among all such cuts, has the minimum value. The maximum flow (max flow)–minimum cut (min cut) theorem states that, for any network with a single source node and a sink node, the maximum feasible flow from the source to the sink equals the minimum cut value for all the cuts of the network.

7.4.3 The cut tree problem

The notion of max flow–min cut for a single source node and a single sink node can be extended to multiple sources and multiple sinks. After finishing a whole *round* for any network according to the maximum flow algorithm explained above, the network is split into two subnetworks. The process is continued with the next round with a different pair of source and sink nodes. Each round is recorded as a level in our computations. The process is continued until all the nodes in the network are given an opportunity to be selected either as a source node or as a sink node in one of the levels. Application of the max flow–min cut notion to the entire network results in what is known as a **cut tree**. A cut tree has a number of properties that are useful in manufacturing decision support systems. The process of efficiently determining the minimum cut between all pairs of nodes in a network to form a cut tree was originally conceptualized by Hu (1969). Their algorithm finds a cut tree by solving iteratively (number of nodes − 1) max flow–min cut algorithms on subnetworks of the original network.

7.4.4 The shortest path problem

The objective in a shortest path problem is to find the shortest path from a source node to a sink node. The edge weights represent the distances. The notion of edge weights as distances can be generalized to activities of any kind, such as activity cost or activity time.

7.4.5 The minimum spanning tree problem

In a minimum spanning tree problem a set of edges that have the shortest total length while providing a path between each pair of nodes is determined. The edges have to be chosen in such a way that the resulting network forms a tree, that is, no cycles are permitted.

7.4.6. Synthesis

In a maximum flow problem the edge capacities act as constraints, whereas in the shortest path and minimum spanning tree problem the edge capacities are not binding constraints, hence they are assumed to be infinite. On the other hand, the edge costs are unimportant in a maximum flow problem, hence every edge is assumed to have a unit cost for convenience. Any other cost distribution which does not distinguish between edges is acceptable for the maximum flow problem. The edge costs are important in shortest path and minimum spanning tree problems; they are used to determine the sets of edges in optimal solutions.

For shortest path and minimum spanning tree problems, a search similar to best-first search is applied from every intermediate node in the search process. For the maximum flow problem a modified version of depth-first search is applied. In a standard depth-first search one backtracks to the last node after a dead end is encountered, whereas in this modified depth first search one backtracks to the root node after reaching a dead end. Note that for the maximum flow problem the root node is the source node and the dead end is the sink node. This modified depth-first search is preferred over the standard depth-first search because the problem structure is better exploited through this process. For example, if one follows the standard depth-first search then one would backtrack to a node immediately preceding the sink node. It becomes computationally more difficult to update the augmenting paths and residual capacities (since one is not storing intermediate states, to save computer memory space) if these values are not systematically recorded beginning from the source node, and this process is better achieved by the modified depth-first search mentioned above.

7.5 THE OBJECT-BASED PROBLEM-SOLVING KNOWLEDGE FOR NETWORK FLOW PROBLEMS

Writing a software code for network flow problems such as determining a cut tree is a typical example of the difficulties encountered in a classical method of coding and illustrates how a library of objects for network flow problems helps in overcoming these difficulties. The versatility of the library of objects is further illustrated by reusing it for solving the representative network flow problems introduced in the previous section. The process is outlined below.

An instance of a *Network* object class is designed to represent each network and its main attributes. Such an instance of a *Network* class is built using a collection of instances of a *Node* object class and a collection of instances of an *Edge* object class. Figure 7.1 shows the *Network* class and its subclasses: *Short-Path, Max-Flow* and *Span-Tree* which are the three network problems used to illustrate the use of the network flow object library in this chapter. The *Cut-Tree* problem is a subclass of the *Max-Flow* problem. The *Short-Path, Max-Flow* and *Span-Tree* objects inherit attributes and services from *Network,* and *Cut-Tree* inherits attributes and services from *Max-Flow*. The modified depth-first search process is implemented as services in *Max-Flow,* and the best-first search is implemented as services in *Short-Path* and *Span-Tree*. While the search logic is the same, the only difference between the *Short-Path* and *Span-Tree* service is in certain testing conditions while applying the best-first search. For example, the starting and stopping conditions for the minimum spanning tree problem and the shortest path problem are different. In a shortest path problem, one attempts to find the shortest path between two nodes or a set of node pairs; in a minimum spanning tree problem one covers the entire network

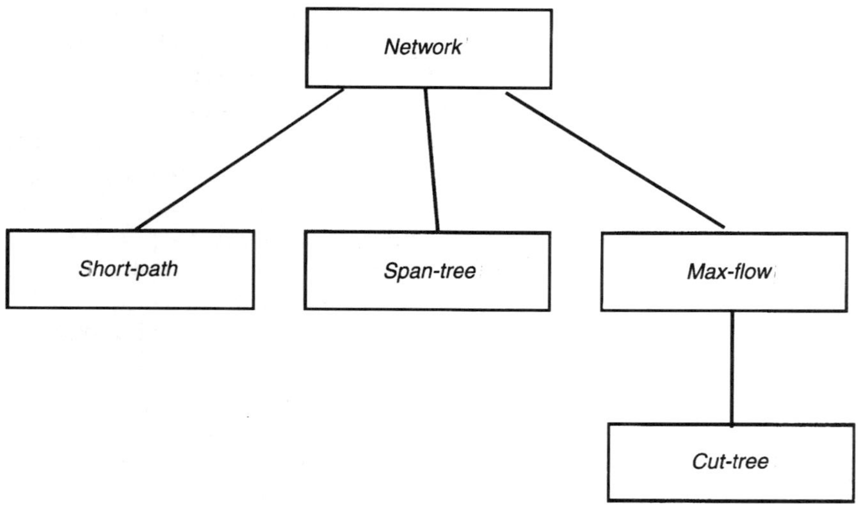

Fig. 7.1 Network class and its subclasses.

without considering any individual pair of nodes.

The attributes of any *Node* are identity, location, number, size, appearance etc. The only attribute needed for solving network flow problems is the identity, in order to define the *Edge* linking any two *Nodes*. The attributes of any *Edge* are its two defining nodes PN1 and PN2. In the *Max-Flow* problem each *Edge* carries a certain flow value, which is referenced by an attribute forwardflow. After each iteration the attribute ForwardFlow is updated. In order to save the original ForwardFlow for building the new network states for the next round, an attribute FlowCapacity is used to save the original data. The *Node* and *Edge* objects are reused for solving the *Short-Path* problem and the *Span-Tree* problem.

The *Tracer* class represents the implementation of the modified depth first search procedure for the max flow problem. Its instances are referred to as *steps*, which represent the traversal of the search space. A *Stack* instance is used to store the *Network* instances generated during the cut tree computation process. Recall that the network is repeatedly divided into two subnetworks during the computation and these subnetworks are stored in the *Stack*. The *Stack* instance is used to store two instances each of the network for the shortest path and the minimum spanning tree problem: one instance indicating the original network and the other instance indicating the last updated network during the search process. The *ControlStack* class has three instances *ControlCut, ControlShort* and *ControlSpan*, and manipulates the stack according to the logic given to it by the *Network* object or one of its subclasses.

7.6 IMPLEMENTATION AND OBSERVATIONS

The object library has been implemented using Borland C++ on an IBM compatible 80486-SX personal computer. The data flows between objects for the maximum flow problem, cut tree problem, shortest path problem and the minimum spanning tree problem are shown in Figs 7.2–7.5. The library has been used to solve a number of network flow problems. The following observations are made during the process of implementing the system.

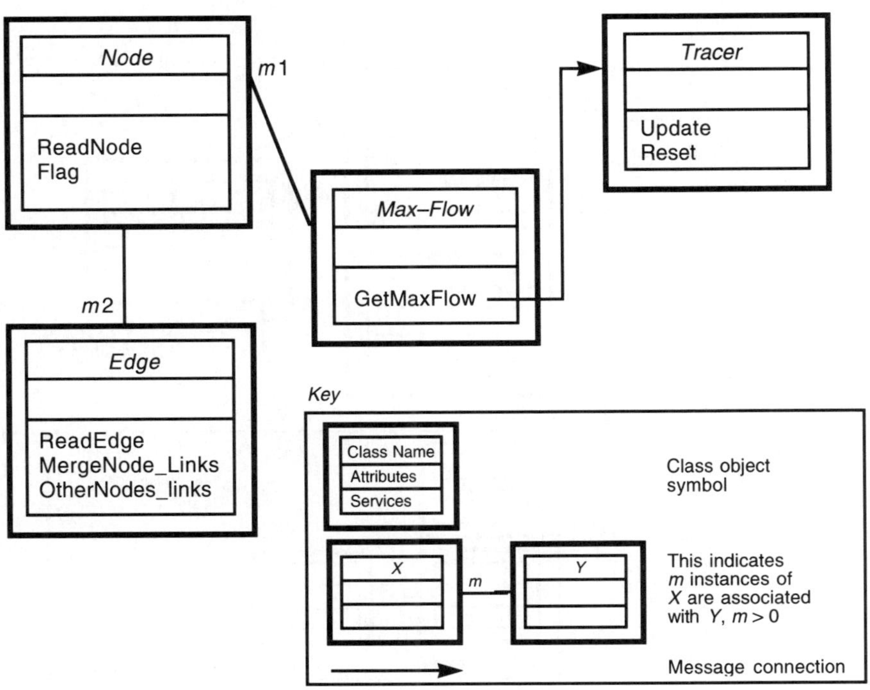

Fig. 7.2 Objects and their connections for the maximum flow problem.

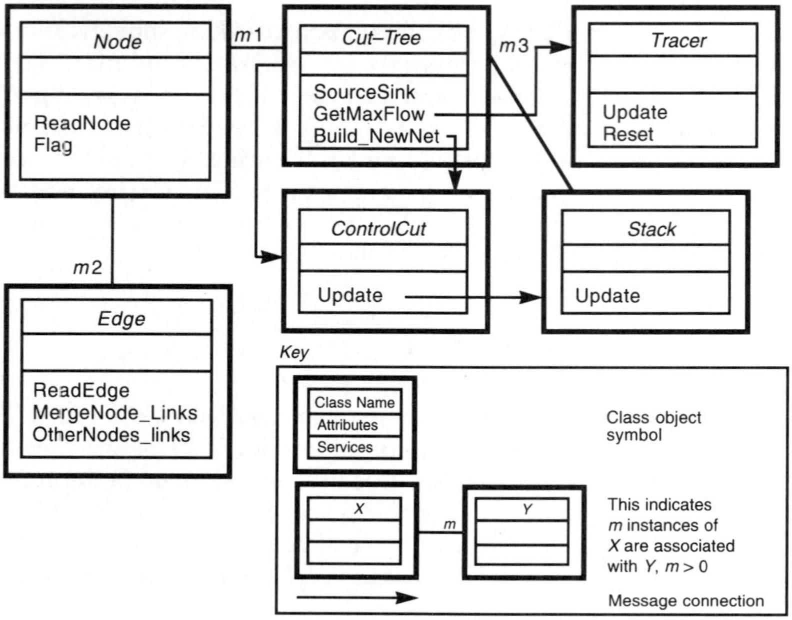

Fig. 7.3 Objects and their connections for the cut tree problem.

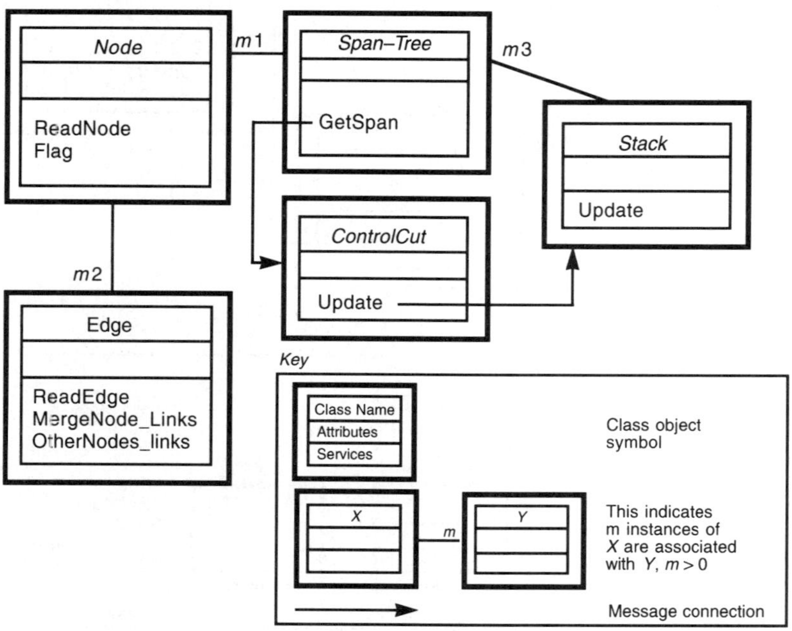

Fig. 7.4 Objects and their connections for the minimum spanning tree problem.

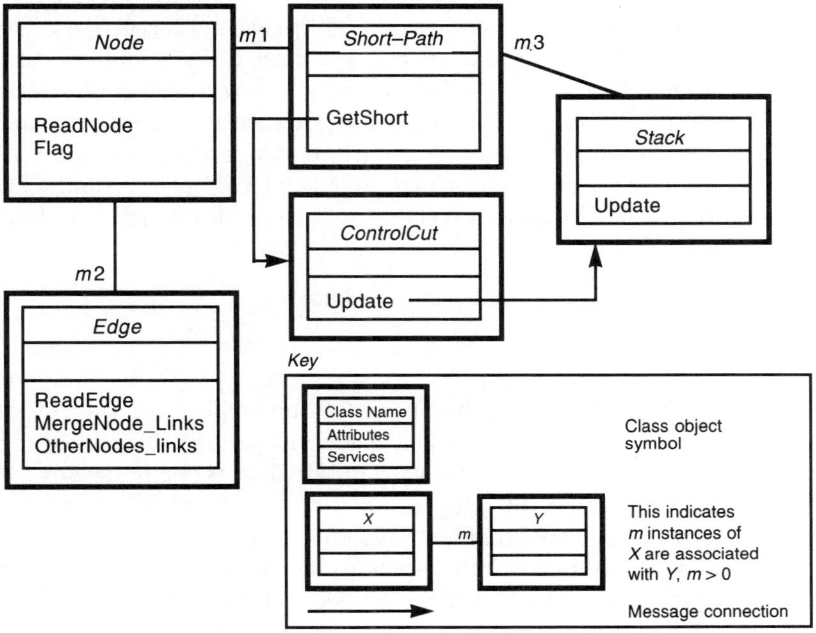

Fig. 7.5 Objects and their connections for the shortest path problem.

7.6.1 Object reusability

Building libraries of routines is the classical approach to reusability; the term 'routine' can be used generically to cover what in the literature is also variously called procedure, function, subroutine, subprogram etc. Normally information about logically connected entities used to be scattered in multiple arrays; hence the scope of reusability used to be very low. To overcome this limitation, libraries of objects are created in an object-oriented technology. Building a library of objects for a class of applications is similar to making software like hardware. For example, just as there are catalogs of VLSI devices, there should be a catalogs of software modules. Just as when a new hardware system is built, components are ordered from catalogs and integrated, so should a software system be a combination of objects from application specific object libraries (Meyer, 1988).

Examples of object reusability for network flow optimization problems are provided in Figs 7.2–7.5. Figure 7.2 shows the object-oriented decomposition of the maximum flow problem. Figure 7.3 shows the object-oriented decomposition of the cut tree problem. Figure 7.4 shows the object-oriented decomposition of the minimum spanning tree problem and Fig. 7.5 shows the same for the shortest path problem. From the four figures, usage of the same objects in different problems is noted, e.g. *Node, Edge, Tracer, Stack, ControlStack* objects get reused. Such objects can be treated as members of an object library for the efficient coding of solutions for network flow optimization problems. In each case certain problem specific features are programmed using the standard objects. Considerable programming time can be saved by object reusability. Also the basis for object reusability represents a step toward a modular software design.

7.6.2 Modularity

One of the key ideas in object-oriented software technology is modularity. Modular programming means the construction of programs from small pieces, usually subroutines. One of the fundamental tenets in achieving modularity is that all information about a module should be private to the module unless it is specifically declared public. This is one of the key ideas behind information hiding (Parnas, 1972), which is essential in manufacturing decision support systems. Modularity reduces the apparent complexity of the initial system description by decomposing it into a set of less complex subsystems connected by a simple structure. For example, in Fig. 7.2 the maximum flow problem is reduced to six less complex modules. A similar pattern is noticed for the other network flow problems shown in Figs. 7.3–7.5.

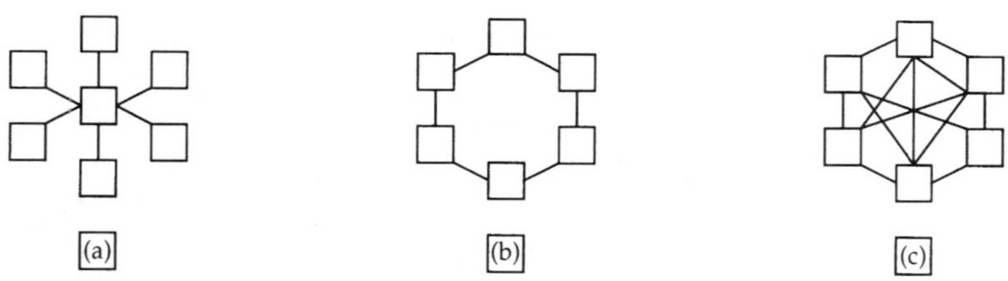

Fig. 7.6 Commonly used type of module interconnection structures.

Another important design strategy is that every module should communicate with as few others as possible. If there are too many relations between modules, then the effect of a change or of an error may propagate to a large number of modules. A small change in problem specification should result in a change of just one module, or a few modules, in a modular system. For the studied problems, the same objects are used, with the only main difference being in the subclasses of the *Network* class as shown in Fig. 7.1.

A third important design strategy is that if any two modules communicate at all, they should exchange as little information as possible. For example, from the various types of module interconnection structure described in (Meyer, 1988) (Fig. 7.6), the one shown in Fig. 7.6(a) corresponds most closely to the modular structure obtained in Figs 7.2–7.5.

7.6.3 Inheritance

Inheritance is another important feature of object-oriented programming. The objects shown in Figs 7.2–7.5 have this feature embedded in them. For example, the service GetMaxFlow in object *Cut-Tree* in Fig. 7.3 is inherited from object *Max-Flow* in Fig. 7.2 in accordance with the inheritance structure shown in Fig. 7.1. Such inheritance further reduces the need for duplication of code and promotes reusability.

7.7 CONCLUSIONS

The described sample of network flow optimization problems is only representative of a very large class of network problems, many of which have considerable similarity with the representative problems described above. The described library of objects can be further generalized to cover a majority of such problems. Such a library can be included as part of a manufacturing decision support software library, somewhere along the lines of mathematical and statistical software libraries.

ACKNOWLEDGEMENTS

This research was supported in part by the NSF under grant DDM 9209849.

REFERENCES

Adiga, S. (1989) Software modelling of manufacturing systems: a case for an object-oriented programming approach. *Annals of Operations Research*, **17**, 363–78.

Bazaraa, M.S., Jarvis, J.J. and Sherali, H.D. (1990) *Linear Programming and Network Flows*, 2nd edn, John Wiley, New York.

Booch, G. (1986) Object-oriented development. *IEEE Transactions on Software Engineering*, **12**(2), 211–221.

Bourne, J.R. 1992 *Object-Oriented Engineering*, Irwin Publishers, Homewood, IL.

Coad, P. and Yourdon, E. (1992) *Object-Oriented Analysis*, Prentice-Hall, Englewood Cliffs, NJ.

Goldberg, A. and Robson, D. (1983) *Smalltalk-80: The Language and its Implementation*, Addison-Wesley, Reading, MA.
Hillier, F. and Lieberman, G. (1990) *Introduction to Operation Research*, 5th edn, McGraw-Hill, New York, pp. 333–358.
Hu, T.C. (1969) *Integer Programming and Network Flows*, Addison-Wesley, Reading, MA, pp. 138–43.
Lenard, M. (1993) An object oriented approach to model management. *Decision Support Systems*, **9**, 67–73.
Lippmann, S.B. (1991) *C++ Primer*, Addison-Wesley, Reading, MA.
Meyer, B. (1988) *Object-Oriented Software Construction*, Prentice-Hall, Englewood Cliffs, NJ.
Montreuil, B. and Banerjee, P. (1988) Object knowledge environment for manufacturing systems layout design. *International Journal of Intelligent Systems*, **3**, 399–410.
Muhanna, W. (1993) An object-oriented framework for model management and DSS development. *Decision Support Systems*, **9**, 217–29.
Parnas, D.L. (1972) A Technique for software module specification with examples. *ACM Communications*, **15**(5), 330–35.
Wilde, N., Matthews, P. and Huitt, R. (1993) Maintaining object-oriented software. *IEEE Software*, January, 75–80.

CHAPTER EIGHT

Machine learning for process parameter selection in intelligent machining

Sanjay S. Jagdale and Tobias Canz

8.1 INTRODUCTION

In a product life-cycle, CAD/CAM integration speeds up the design and manufacturing process of a product significantly. Despite the fact that the correct process parameters have significant influence on the quality of the final product, few attempts have been made to incorporate the systematic selection of the process parameters of a metal-cutting operation in a facility-specific manner. It is believed that incorporating the determination of machine-specific process parameters in the process planning phase will help to further increase productivity in the manufacturing phase of the product cycle. In addition, it is believed that the quality of products will improve, and costs for follow-up treatment of parts produced in a cutting process will decrease. Ideally, the evaluation criteria for a machining operation (such as surface finish, accuracy, cutting forces and temperature at tool tip) should be controlled in a closed-loop control system. The controller of such a system would use the deviation of the measured values of the evaluation criteria (as produced by the process) from their desired values and would then produce a control signal to adjust the process parameters. Narendra and Parthasarathy (1990) propose the use of neural networks for the intelligent controller of an arbitrary system which does not require an underlying mathematical model of the controlled process. For such an imaginary system, the determination of initial process parameters via approximate machining tables would be adequate because the process parameters would be changed in the course of the machining operation. However, sensor accuracy and equipment limitations do not allow the use of this method for machining operations. The consequence is that the highly dynamic process of machining can only be influenced by the choice of the initial process parameters that are changed solely based on geometry (in NC machining). Acquisition of measured data containing information regarding the process parameters and the achieved evaluation criteria makes it feasible to capture machine-specific behavior for the determination of initial process parameters. Machine learning techniques are used as tools that learn to map the process parameters to a set of evaluation criteria from sampled training cases. In this study, a simulation program was used to generate training and test cases. Various neural network structures with different numbers of hidden units mapped the process parameters to the evaluation criteria.

8.2 MACHINE LEARNING TECHNIQUE

The neural network algorithm chosen was a feedforward algorithm that updates the weights between the layers of the network in a back-propagation step. The network incorporates one hidden layer and a variable number of units in each layer. The input layer is referred to as vector $\boldsymbol{x}$, the hidden layer as vector $\boldsymbol{y}$ and the output layer as vector $\boldsymbol{z}$. Weights connecting each node of a previous layer with a node of the next layer are stored in the matrix $\boldsymbol{V}$ for connecting the input layer with the hidden layer and in matrix $\boldsymbol{W}$ for connecting the hidden layer with the output layer. The network is designed so that the mean square error (MSE) is propagated backward through the network after the presentation of all patterns in the training set. One presentation of all patterns is called a sweep. During a single sweep, all partial derivatives of the error with respect to the weights in the two matrices are summed up and the sum is used to update the weights in the corresponding weight matrix. By means of this procedure, the network can learn to map a set of inputs to a set of outputs. Learning is carried out by iteratively adjusting the coupling strengths in the network so as to minimize the differences between the actual output state vector $\boldsymbol{z}$ of the network and the target state vector $\boldsymbol{t}$. The updating of the weights is repeated until either the error-measure falls below a desired threshold or a maximum number of sweeps has been trained. Updating of the bias vector values is done in a similar fashion. In the present application (machining parameter selection), the procedure can be referred to as an inverse mapping problem because the desired output of the machining process is presented at the input nodes of the network and the required process parameters (input to the machining process) are deduced at the output layers.

8.3 DATA ACQUISITION

Data points for training and test cases were generated by using a simulation tool for orthogonal turning operations (Polisetty, 1992).

Simulation was carried out for a workpiece material of 0.2% low carbon steel. It was assumed that machining is done with a cutting-grade tungsten-carbide (WC–TiC–TaC–Co) cutting tool; the tool has a rake angle of 5° and a groove-type chip breaker. Starting tool and workpiece temperatures of 25°C were used. Table 8.1 shows the ranges of the process parameters which were used as inputs to the simulation program. In order to consider only feasible machining set-ups, the process parameters were classified into several ranges. Within these ranges, the combination was assumed to be random. The choice of the ranges for the process parameters was guided by Parter, Lascoe and Nelson (1967).

Table 8.1 Process parameter ranges

Range	*Feed Rate (mm/rev)*		*Depth of cut (mm)*		*Cutting speed (m/s)*	
	min.	max.	min.	max.	min.	max.
1	0.05	0.09	0.1	0.2	4.8	6.1
2	0.09	0.13	0.2	0.3	3.6	4.8
3	0.13	0.25	0.3	1.4	3.2	3.6
4	0.25	0.38	1.4	2.3	2.8	3.2
5	0.38	0.58	2.3	3.5	2.4	2.8
6	0.58	0.76	3.5	4.7	2.0	2.4

Table 8.2 Simulation results

Evaluation criteria	*Minimum value*	*Maximum value*
F_c(N)	11.4	5289.4
F_t (N)	8.5	2841.0
W_{ad} (mm/s)	0.220243	0.664287
W_{dif} (mm/s)	0.000001	0.012802
Chip quality	1	3
Finish (μm)	7.4	41.8
Temperature (°C)	895	1296

In the data set that was used, the number of training and testing patterns was 1936 and 888, respectively. These numbers closely match the recommended ratio of training to testing patterns which should be 2 to 1 (Weiss and Kalikowski, 1991). Table 8.2 shows the minimum and maximum values of the evaluation criteria resulting from the simulation of the orthogonal turning process. F_c and F_t are the cutting and thrust forces, respectively; W_{ad} and W_{dif} are measures of the adhesive and diffusive tool wear. The other three evaluation criteria listed are the chip quality, the surface finish and the temperature at the tool tip.

8.3.1 Statistical analysis of the simulation data

Rank order correlation coefficients were used for the determination of correlation between process parameters and evaluation criteria. Kendall's τ or Spearman rank order correlation coefficients can be applied to data that are not quantitative or when a normal distribution cannot be assumed, 1977). The Spearman correlation coefficients were used in this study. The correlation factor between two variables x and y is calculated using the ranked value of these variables. The variables are ranked separately according to their values. For a pair i of x and y_i, the Spearman rank order correlation coefficient can be determined from the correlation coefficient for normally distributed data by replacing variable values with their ranks. For a data set of N samples, the rank order correlation coefficients are as follows:

$$r_s = 1 - \frac{6\sum_{i=1}^{N} d_i^2}{N(N^2 - 1)} \tag{8.1}$$

where d_i is the difference in rank between x_i and y_i (Romano, 1977). An r_s value of +1 indicates perfect agreement in the ranks of the x and y values. An r_s value of −1 shows that the x-values and the corresponding y-values are ranked in the exact opposite order. The results of this analysis are presented in the matrix in Fig. 8.1.

	F_c	F_t	W_{ad}	W_{dif}	Chip quality	Surface finish	Temperature
Feed rate	0.9873648	0.9828569	-0.9716635	0.8320603	0.8660184	0.9999356	0.4960235
Depth of cut	0.9939075	0.9949245	-0.9713830	0.8915256	0.8526623	0.9728374	0.4559548
Cuttingspeed	-0.9734570	-0.9778443	0.9998978	-0.7903215	-0.8513157	-0.9706479	-0.4089008

Fig. 8.1 Correlation matrix for the process parameters and the evolution criteria.

Similar correlation matrices were obtained for the analysis of the correlation among the process parameters and among the evaluation criteria. Due to the strong correlation between the process parameters and the evaluation criteria, a complete set of process parameters is predicted from a set of evaluation criteria as opposed to earlier studies. Earlier studies (e.g. Chryssolouris and Guillot, 1988) that focused on the modeling of the machining process predicted one evaluation criterion at a time from a set of process parameters. It was claimed that reversing this approach would help in predicting process parameters from the desired evaluation criteria. Neglecting the correlation of process parameters and evaluation criteria as well as the correlation among the evaluation criteria and among the process parameters themselves results in a loss of information. The consequences are inaccuracies in the model of the machining process and the inadequacy of the inverse model proposed for the prediction of the process parameters. The following dominance discussion makes this point clear.

If it is observed that as process parameter 1 ($pp1$) increases, evaluation criteria 1 (ec_1) also increases, and if it is further true that pp_2 has a similar effect on ec_1, then can this observation be generalized. The question would be the following. If $pp^a{}_1 < pp^b{}_1$ (meaning process parameter 1 in instance a is smaller than the same process parameter 1 in instance b) and if $pp^a{}_2 < pp^b{}_2$, does it necessarily mean that $ec^a{}_1 < ec^b{}_1$ This first proposition can be written as follows.

Proposition 8.1

Let

$$pp_1 \uparrow \Rightarrow ec_1 \uparrow$$
$$pp_2 \uparrow \Rightarrow ec_1 \uparrow .$$

Then if

$$pp_1^a < pp_1^b \text{ and } pp_2^a < pp_2^b$$

does it imply that

$$ec_1^a < ec_1^b?$$

The question is whether pair b dominates pair a. A related issue can be expressed as follows.

Proposition 8.2

Let

$$pp_3 \uparrow \Rightarrow ec_2 \uparrow$$
$$pp_4 \downarrow \Rightarrow ec_2 \uparrow .$$

Then if

$$pp_3^c < pp_3^d \text{ and } pp_4^c < pp_4^d$$

does it imply that

$$ec_2^c < ec_2^d?$$

Only if both of the propositions above are true, would the 'one-process-parameter-at-a-time' approach be correct for the prediction of the process parameters. The approach introduced in this chapter will yield correct results whether the propositions are true or not. Hence it is applicable to more general cases. The rank order correlation coefficients, as introduced above, implement the dominance principle. A pair of dominating and dominated variables have a rank order correlation coefficient of +1 or −1. As can be seen in the correlation matrix presented in Fig. 8.1, this condition is not the case for the data used in this study.

8.4 RESULTS

The following initial values were used for the training of all networks evaluated. A learning rate η of 0.00465, together with a momentum α of zero were chosen. The entries in the bias vectors were set to 1. The weight matrices $\boldsymbol{V}$ and $\boldsymbol{W}$ were initialized using random values between 0.0001 and 0.01.

8.4.1 Mean square error of the training cases

The mean square error (MSE) is defined as follows:

$$MSE = \frac{1}{2}\sum_{c=1}^{N_c}\sum_{j=1}^{N_z}(z_{j,c} - t_{j,c})^2 \tag{8.2}$$

The MSE of the training patterns in the training phase is used during training of the network to stop the training phase if the MSE falls below a given maximum. Table 8.3 shows the MSE achieved with different networks depending on the number of iterations.

It can be seen that the MSE for each of these networks was reduced by a factor of about 150 which is well within the expected range. Table 8.3 shows that the lowest MSE is achieved for a 7–5–3 network followed by the 7–2–3 and 7–3–3 networks. All networks perform similarly well on the training patterns. The graph in Fig. 8.2 shows a plot of the MSE versus the number of iterations.

Table 8.3 Mean square error as function of training sweeps

Iterations	*7–2–3 network*	*7–3–3 network*	*7–5–3 network*
1	0.088793	0.088000	0.089320
15	0.023001	0.023114	0.023754
105	0.002213	0.002155	0.002160
525	0.001695	0.001707	0.001770
1 005	0.001178	0.001265	0.001430
2 010	0.000711	0.000780	0.000902
3 015	0.000644	0.000694	0.000735
4 020	0.000611	0.000663	0.000676
4 995	0.000592	0.000635	0.000617
6 000	0.000578	0.000616	0.000569
6 990	0.000568	0.000603	0.000544
8 000	0.000561	0.000593	0.000531
9 005	0.000555	0.000585	0.000523
9 995	0.000551	0.000578	0.000520
11 000	0.000547	0.000573	0.000520

8.4.2 Maximum and average errors for the test patterns

Another performance measure is the maximum error for each of the variables predicted in the testing phase. Table 8.4 summarizes the average and maximum errors, as a percentage, for each of the predicted process parameters depending on the network structures.

The 7–3–3 network seems to be best suited for predicting the feed rate of the cutting operation. The 7–2–3 network performs best for the other two process parameters if the objective is to reduce the maximum error. Due to its lowest bias in the error vector, the 7–5–3 network is capable of predicting the depth of cut and cutting speed with the lowest average error.

8.4.3 Residual noise in the error signal

The gradient descent method does not stop 'learning' until the error vector is zero. If, for the test patterns, too much of the error was contributed by a non-random error, then it can be said that the training was not successful. On the other hand, when trying to further minimize the MSE, if the error is caused by random noise only, then increases in the computational effort were experienced without improving the performance of the network. The basic idea for testing the presence of random noise in the error vector of the testing phase can be stated in the following

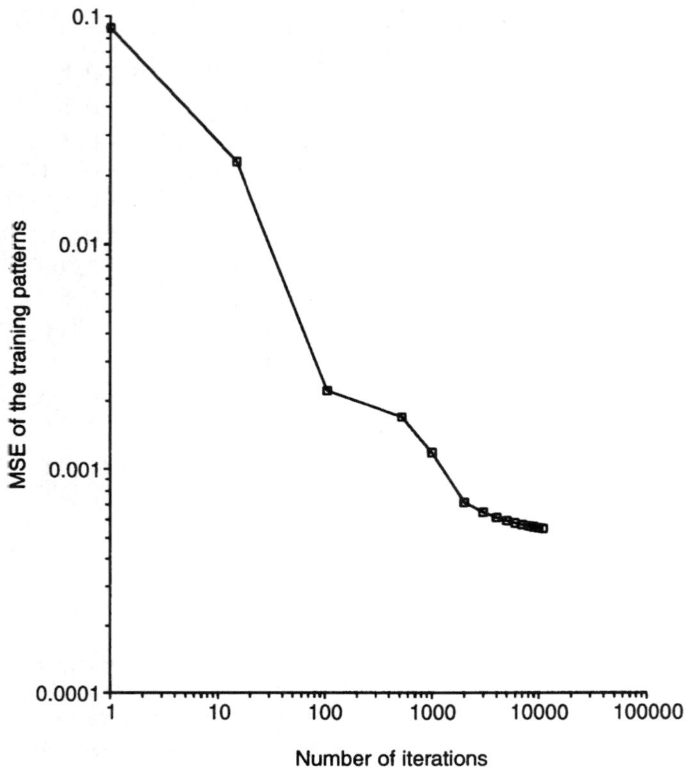

Fig. 8.2 Mean square error for training the 7–2–3 net.

way. If two variables are independent, the covariance measure and the correlation measure between the two variables are zero.

The complex n-dimensional random variables compared are called $\boldsymbol{x}$ and $\boldsymbol{y}$ and are represented as row vectors. The expression $E\{\boldsymbol{x}\}$ denotes the expectation of variable $\boldsymbol{x}$ and is equivalent to $\sum_{i=1}^{N} X_i$. The cross-correlation matrix between the two vectors is defined as follows:

$$\boldsymbol{R} = E\{\ \boldsymbol{x}^T\ \boldsymbol{y}^*\} \tag{8.3}$$

If either $E\{\boldsymbol{x}\}$ or $E\{\boldsymbol{y}\}$ is zero, the cross-correlation matrix yields an unnormalized correlation matrix (assuming linearity) (Papoulis, 1984). For this problem, the cross-correlation matrix can be interpreted in the following way. The error vector $\boldsymbol{r}$ and the target vector $\boldsymbol{t}$ are uncorrelated if all entries in the cross-correlation matrix are zero. In such a case, the

Table 8.4 Summary of the performance of the neural network structures

		Error (%)	
Network structure	*Feed rate*	*Depth of cut*	*Cutting speed*
7–2–3	Max. 11.72	14.04	8.55
	Ave. 3.14	3.74	1.64
7–3–3	Max. 10.70	14.81	9.98
	Ave. 3.08	3.67	2.07
7–5–3	Max. 15.12	14.35	8.71
	Ave. 3.45	2.99	1.37

network does not produce a different error vector ***r*** when trying to achieve another target vector ***t*** and learning has found an optimal solution in which the residual error is due to noise.

As can be seen from Fig. 8.3(a) and 8.3(b), the correlation between the components of the vectors is reduced by a factor of at least 25 (and, sometimes, up to a factor of over 280). This indicates that learning has significantly reduced the correlation between these two vectors.

	t1	t2	t3
r.1	-0.066881	-0.070040	0.036149
r.2	-0.070040	-0.078524	0.039056
r.3	0.036150	0.039057	-0.027638

(a)

	t2	t3	t3
r.1	-0.001215	0.000627	-0.000277
r.2	0.001940	-0.000404	-0.001576
r.3	0.000128	-0.000185	-0.000529

(b)

Fig. 8.3 Cross-correlation matrix for (a) the untrained network; (b) the trained network.

8.4.4 Bias of the error vector

The components of the bias vector can help to identify if the mean values of the entries in the error vector ***r*** are zero. A bias of zero indicates that errors in the prediction are distributed equally around a zero error. This condition means that the results are not biased to either the high side or the low side of the target vector. For N testing patterns, the entries b_i in the bias vector ***b*** are calculated as follows:

$$b_i = \frac{1}{N}\sum_{i=1}^{N} r_i \qquad (8.4)$$

Table 8.5 Bias vectors for different networks

Component	*Variable*	*Trained 7–2–3*	*Trained 7–3–3*	*Trained 7–5–3*	*Untrained 7–2–3*
1	Feed rate	0.000797	0.000699	0.000396	0.385866
2	Depth of cut	−0.003706	−0.002261	0.000639	0.420898
3	Cutting speed	−0.001641	−0.002251	−0.000839	0.204609

Table 8.5 lists the entries in the bias vectors for the trained versions of the 7–2–3, 7–3–3 and 7–5–3 networks. To contrast the difference with respect to an untrained network, the last column of the table shows the bias vector for the untrained 7–2–3 network.

The results show that the bias vectors look similar for the 7–2–3 and 7–3–3 trained networks. The 7–5–3 network performs significantly better for this performance measure, causing the network to produce the lowest average error for two of the three process parameters predicted in the test phase. A comparison of the values for the untrained and trained versions of the 7–2–2 network shows that the improvement in the bias is by a factor that ranges from 100 to about 500. This is well within the expected range and shows that the network works correctly.

8.4.5 Computation time for training of the network

Approximate training times for networks of different sizes are given in Table 8.6. The table shows the CPU times in seconds for runs on VAX 4000/500 and Sun-4/65 workstations. The operation mode chosen includes training a network from scratch and testing the results after 100 iterations of the network. The CPU times for the VAX and the Sun workstations should not be mutually compared because they have been acquired in two different ways. They are intended merely as a reference for readers who run similar programs on the same type of equipment. On the VAX workstation, CPU times were determined as the time needed for computations only; I/O times were not included.

Table 8.6 CPU times in seconds for 100 iterations

Network	*Sun-4/65 (s)*	*VAX 4000/500 (s)*
7–2–3	1220	173
7–3–3	1317	215
7–5–3	1468	273
7–9–3	1799	367

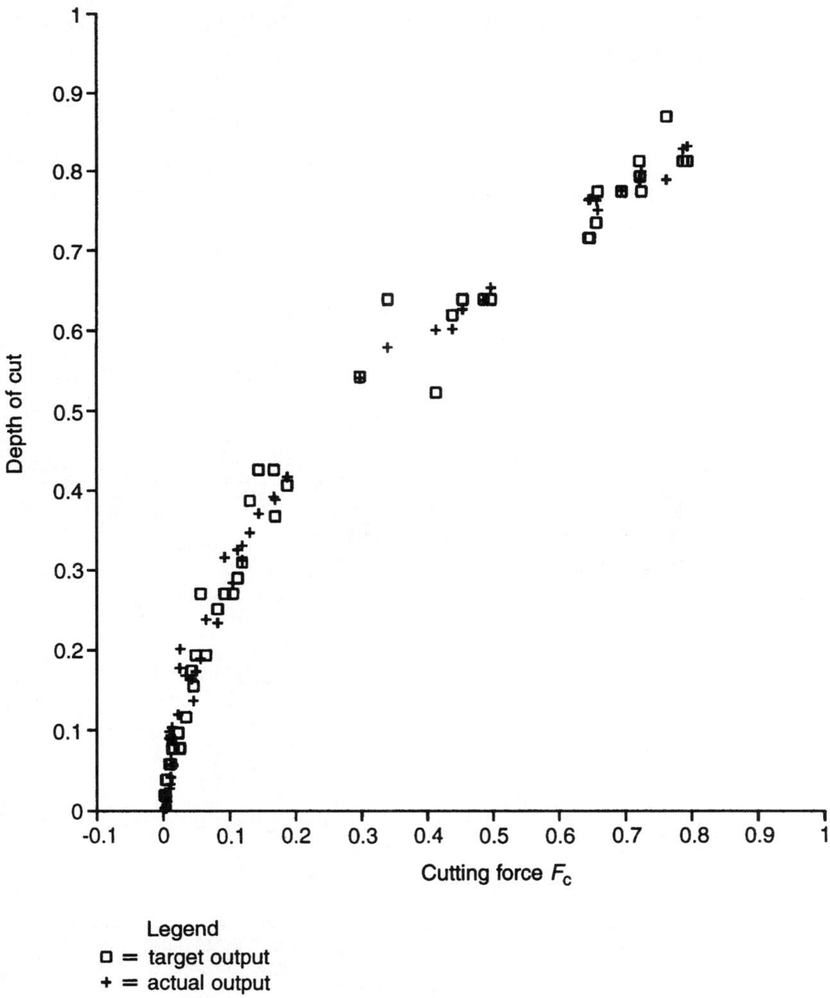

Fig. 8.4 Results for depth of cut and cutting force.

8.4.6 Graphs of network performance

All the graphs show normalized data that were used for the training of the neural network. Since the maximum value for all variables has been normalized to 0.91, an error of 0.091 in these graphs equals an error of approximately 10%. Figures 8.4–8.15 show the performance of the 7–2–3 network (11 000 iterations). Other networks produced graphs that look very similar. Figures 8.4 and 8.5 use the cutting force F_c as variable. In Figs 8.6–8.9 the adhesive tool wear is used. The graphs in Figs 8.10 and 8.11 show the results for the tool wear due to diffusion. Finally, Figures 8.12–8.15 employ the surface finish as the variable. Figure 8.10 clearly shows the advantage of using neural networks for the mapping of data that does not exhibit any regular trend or pattern.

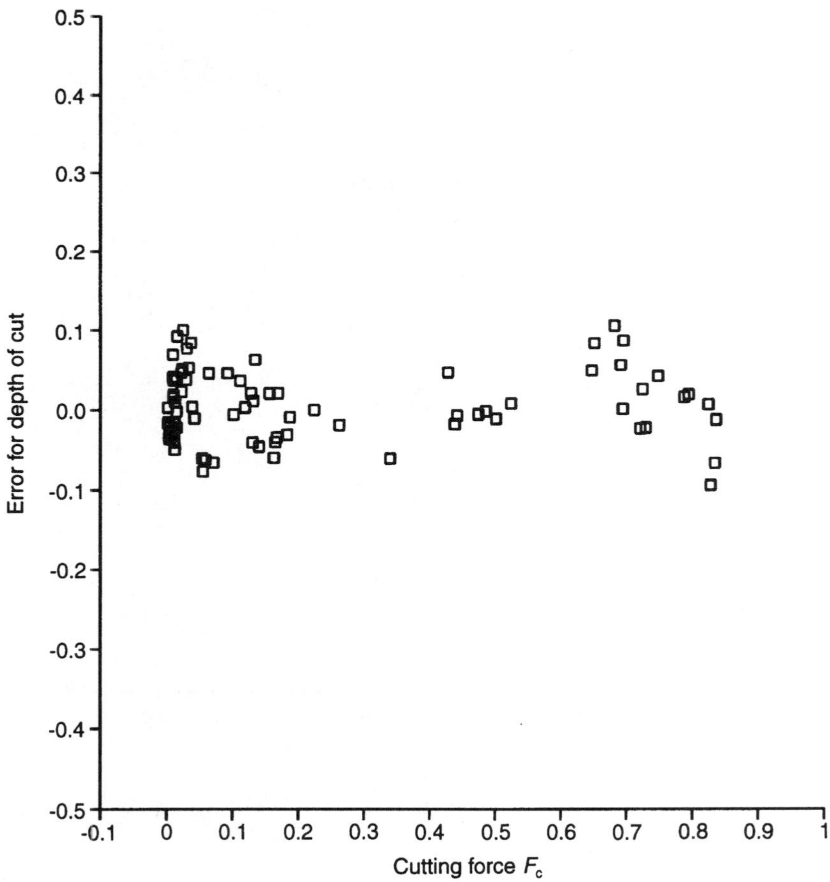

Fig. 8.5 Error for depth of cut and cutting force.

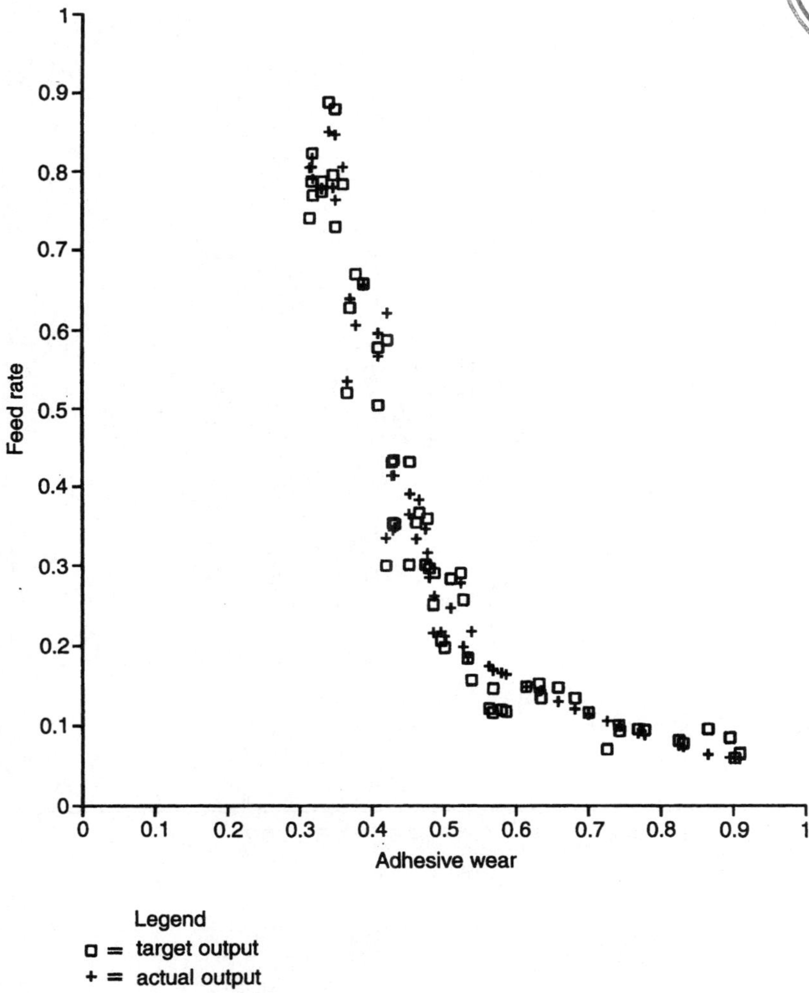

Fig. 8.6 Results for feed rate and adhesive wear.

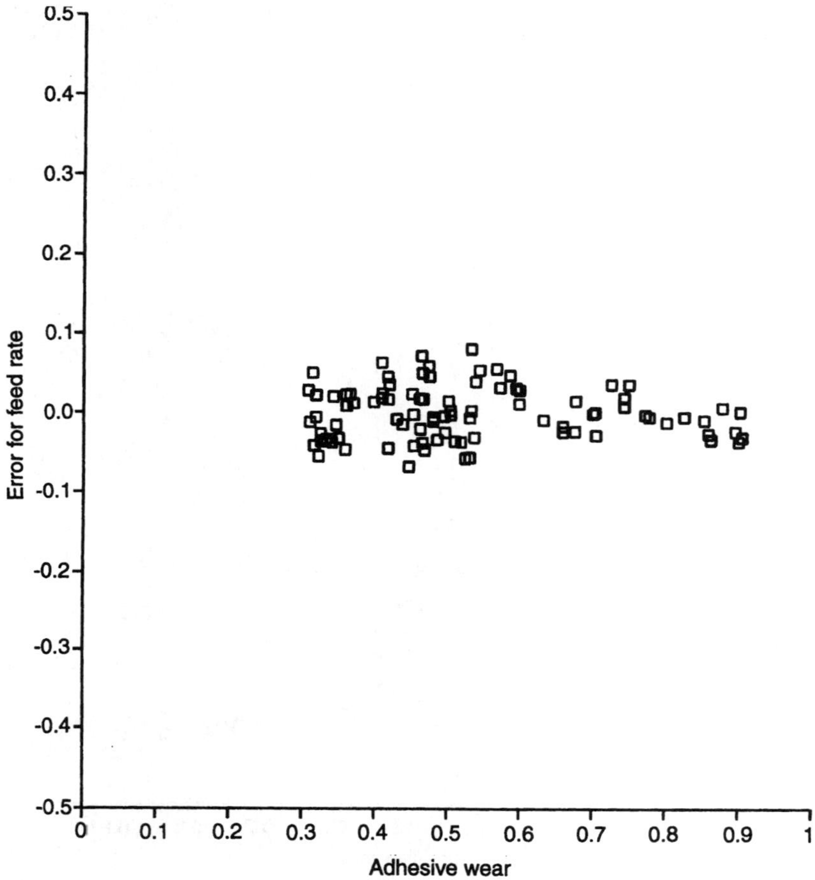

Fig. 8.7 Error for feed rate and adhesive wear.

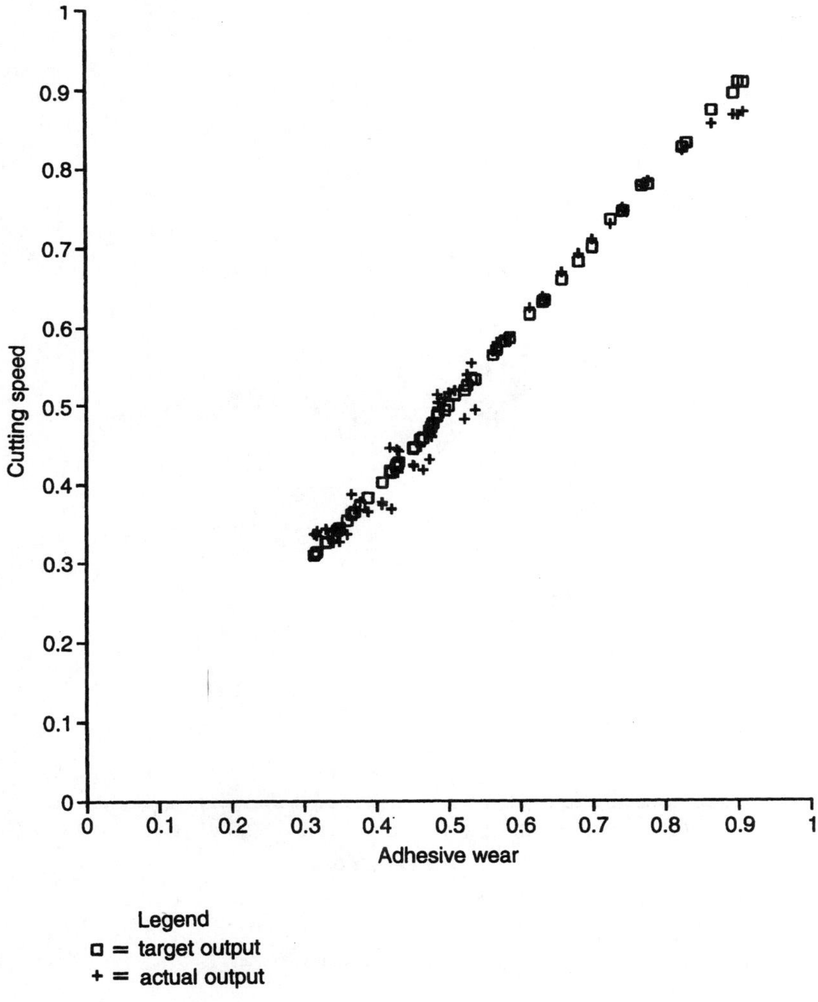

Fig. 8.8 Results for cutting speed and adhesive wear.

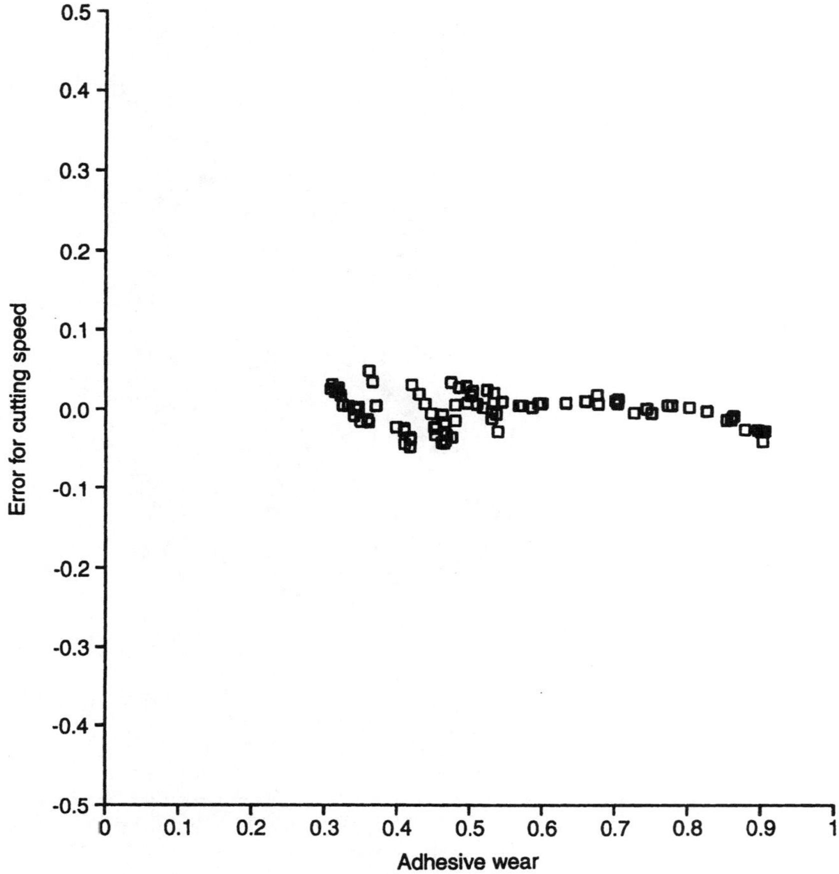

Fig. 8.9 Error for cutting speed and adhesive wear.

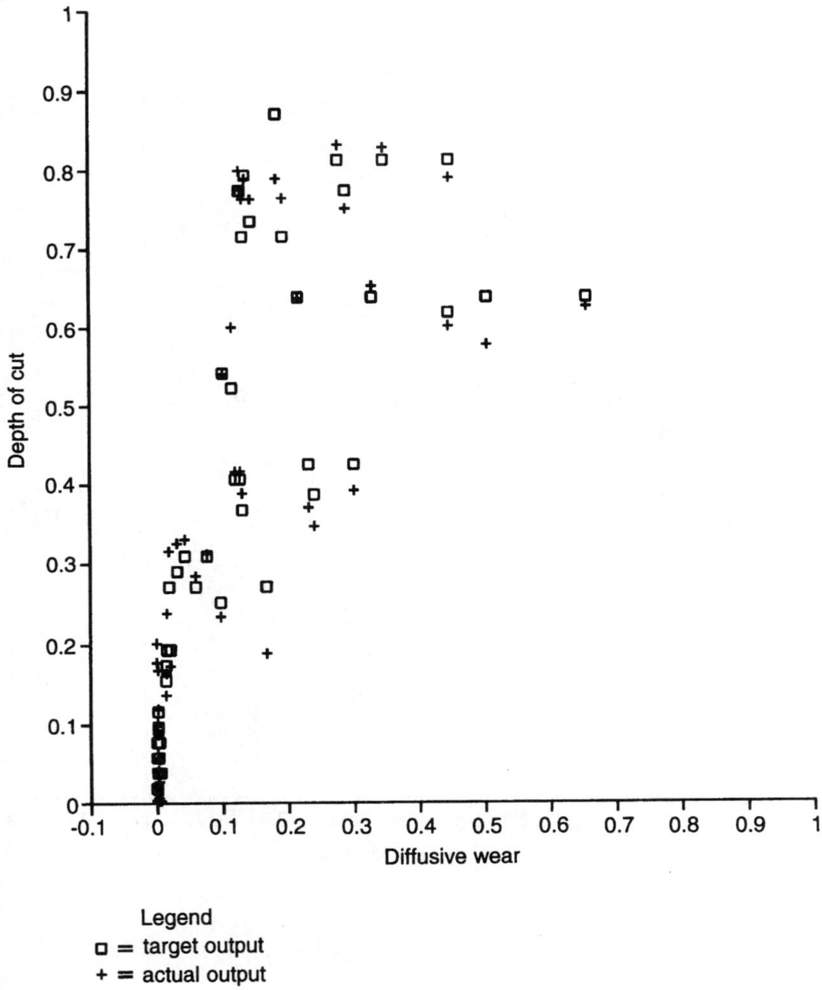

Fig. 8.10 Results for depth of cut and diffusive wear.

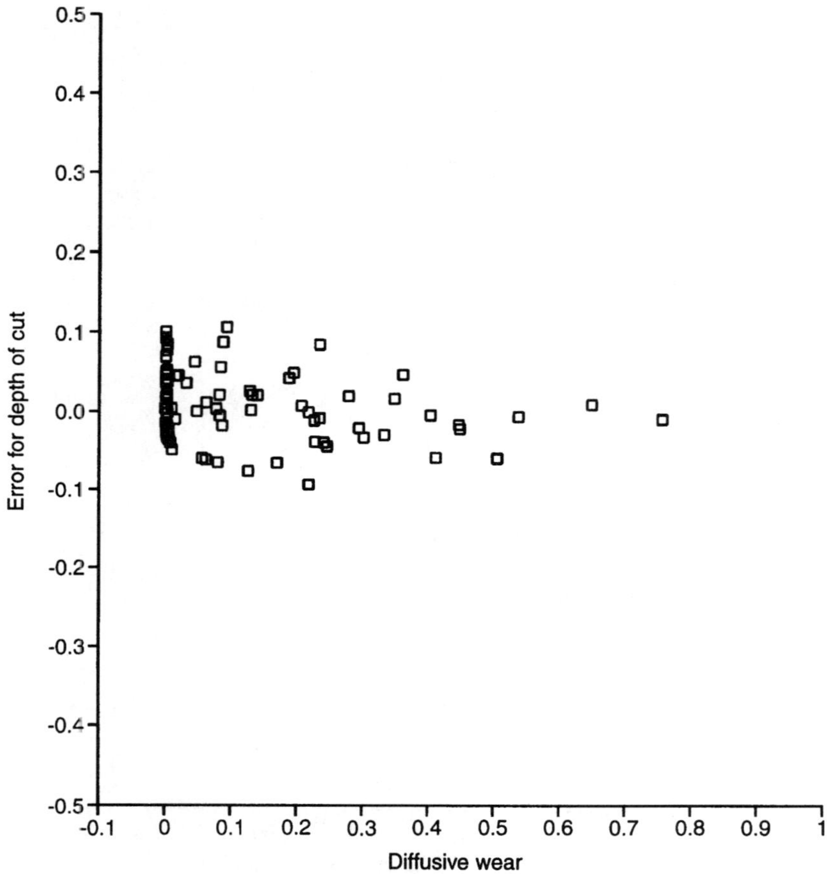

Fig. 8.11 Error for depth of cut and diffusive wear.

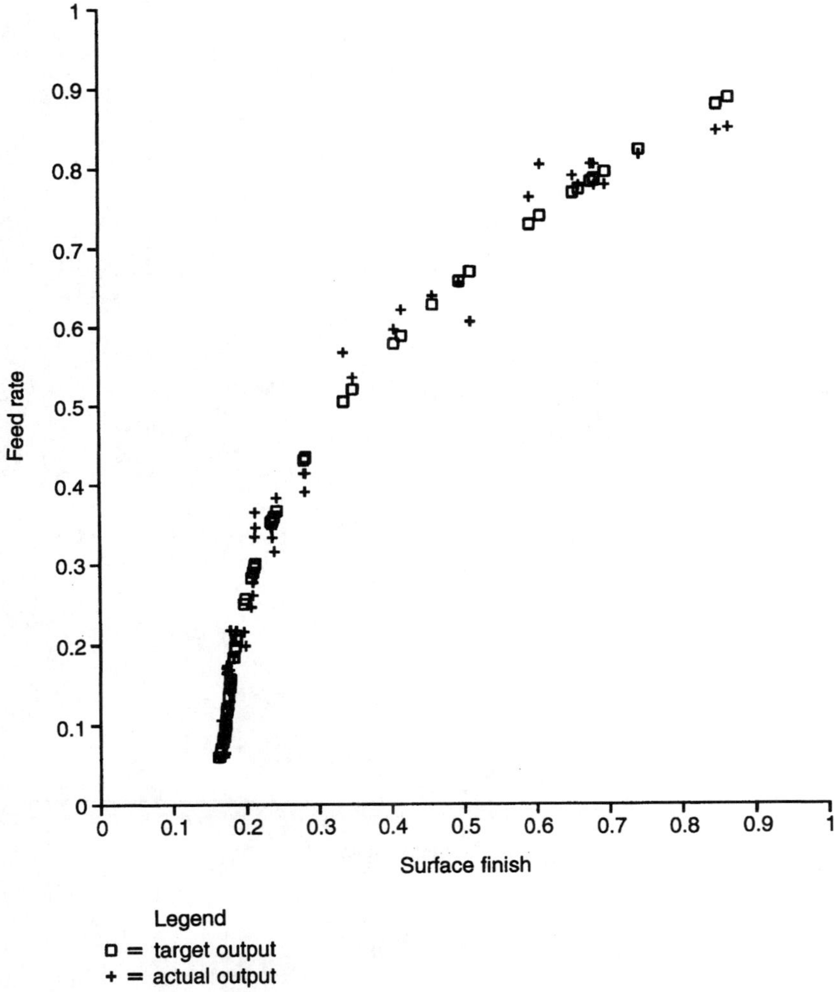

Fig. 8.12 Results for feed rate and surface finish.

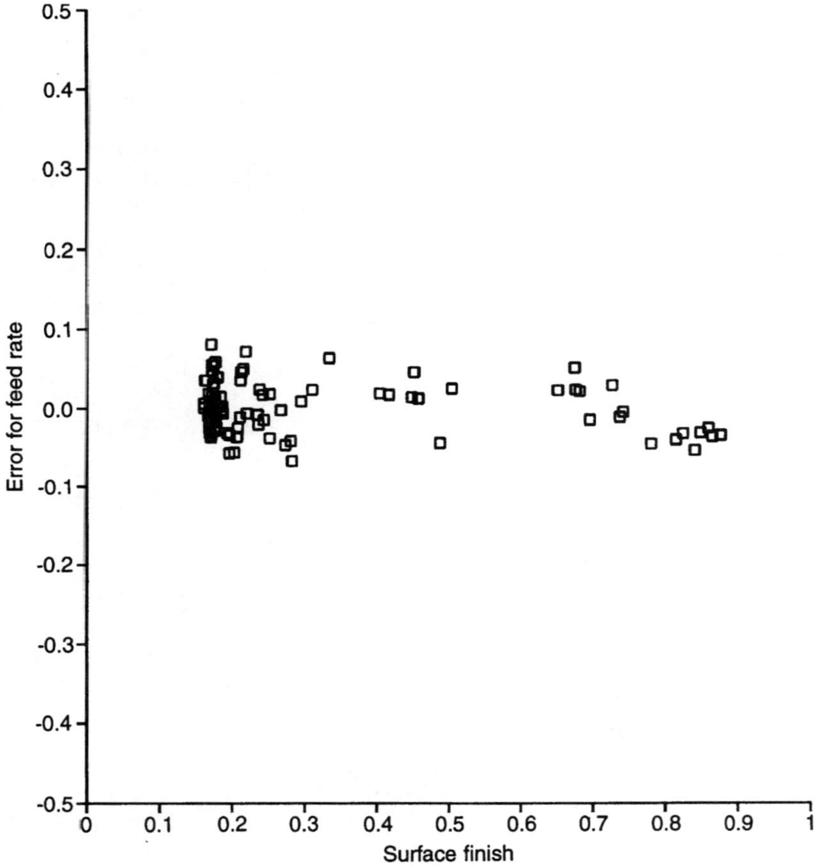

Fig. 8.13 Error for feed rate and surface finish.

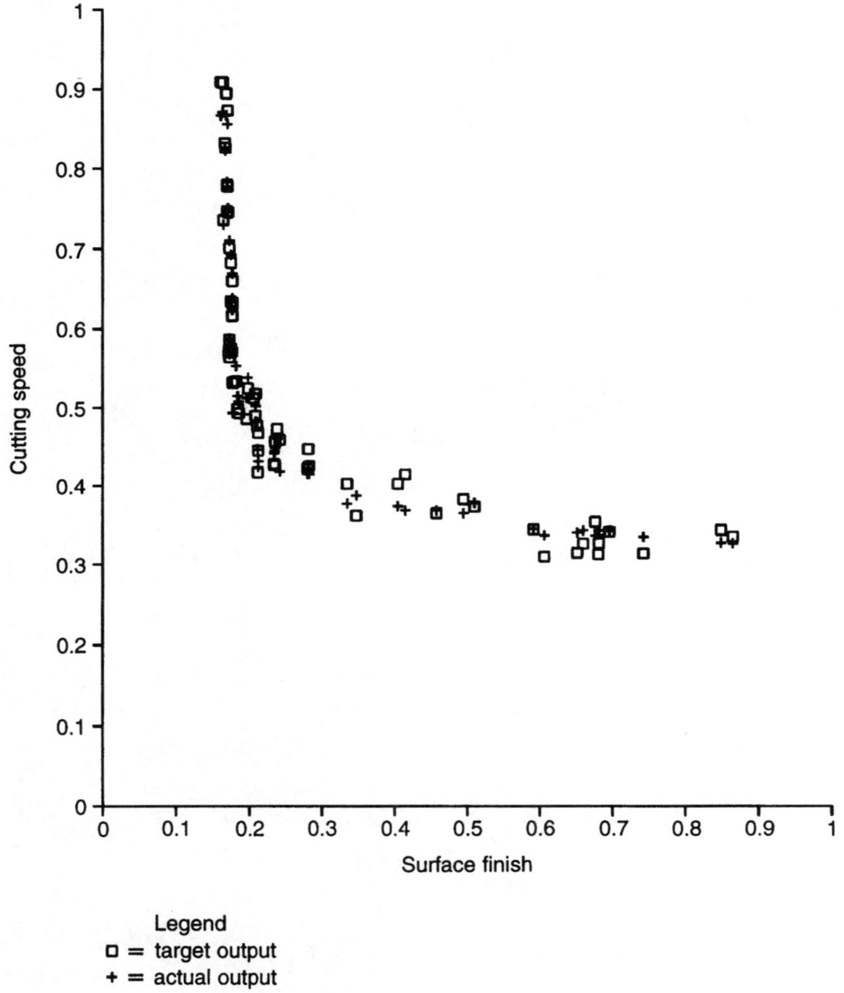

Fig. 8.14 Results for cutting speed and surface finish.

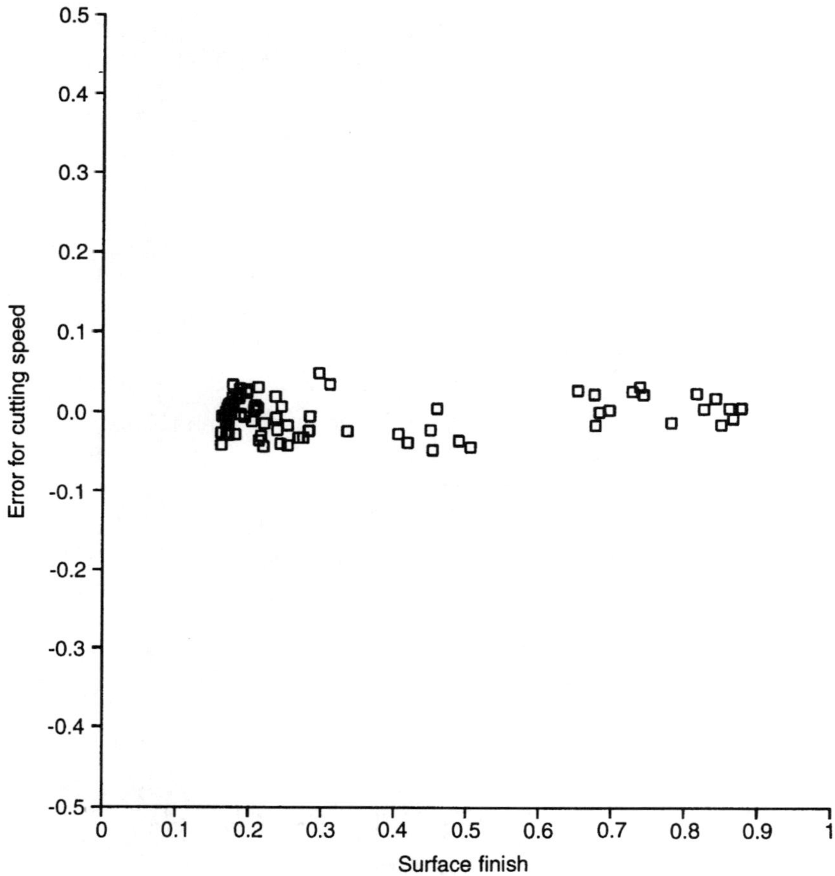

Fig. 8.15 Error for cutting speed and surface finish.

8.5 CONCLUSIONS

This study focused on machine learning in the decision-making process concerning process parameter selection for an orthogonal cutting operation on a lathe. The intent was to show that process parameters for a specific machine tool can be predicted from the evaluation criteria of the cutting operation by the use of neural networks. Process parameters chosen were feed rate, depth of cut and cutting speed. Inputs to the neural networks were the evaluation criteria, namely, the cutting forces, tool wear, chip quality, surface finish and cutting temperature. Emphasis in this study was put on dominance-related issues. The study showed that handling dominance-related issues allows the prediction of process parameters from evaluation criteria with simpler neural networks more accurately than in previous studies. All three networks that were used performed to expectation. However, networks with more than five nodes in the hidden layer did not converge. The final decision regarding which network structure to use should be determined from machine-specific performance data.

REFERENCES

Chryssolouris, G. and Guillot, M. (1988) An A.I. approach to the selection of process parameters in intelligent machining, in *ASME Proceedings of the Symposium on Sensors and Controls for Manufacturing*, Chicago, IL, 27 November–2 December, pp. 199–206.

Narendra, K.S. and Parthasarathy, K. (1990) Identification and control of dynamical systems using neural networks. *IEEE Transactions on Neural Networks*, **1** (1), 4–27.

Papoulis, A. (1984) *Probability, Random Variables, Stochastic Processes*, McGraw-Hill, New York.

Parter, H.W., Lascoe, O.D. and Nelson, C.A. (1967) *Machine Shop Operations and Setups*, American Technical Society, Chicago, IL.

Polisetty, S. (1992) *Simulation and performance evaluation of metal cutting processes from NC programs*. Master's Thesis, University of Arizona, Tucson, AZ.

Romano, A. (1977) *Applied Statistics for Science and Industry*, Allyn and Bacon, Boston, MA.

Weiss, S.M. and Kulikowski, C.A. (1991) *Computer Systems That Learn*, Morgan Kaufmann Publishers, San Mateo, CA.

CHAPTER NINE

An interactive program for machine grouping and layout

Sunderesh S. Heragu and Sivakumar R. Kakuturi

9.1 INTRODUCTION

A number of companies have reaped the benefits offered by cellular manufacturing systems (CMSs) (Wemmerlov and Hyer, 1986). Typical benefits include: set-up time reduction, improved machine utilization, an increase in production rate and employee morale, a decrease in labor costs and traffic congestion and better control of the manufacturing system. Cellular manufacturing (CM) involves grouping of machines based on the parts they manufacture. The grouping of machines and parts in a CMS is briefly explained here by considering a simple example. Figure 9.1 shows four part routings. The first row indicates that part 1 is routed through machines A, C and E in succession. Figure 9.2 shows the arrangement of the machines referred to in Fig. 9.1 in a job shop environment. Figure 9.3 shows the arrangement of the same machines in 'cells'. Notice that each cell is dedicated to production of a set of parts. The main idea behind grouping 'like' machines into cells is to minimize intercellular part movement. After grouping is done, the next logical step is to develop a layout for the cells and machines within each cell so that both intercellular *and* intracellular material handling cost is minimized. This step, however, is often ignored. Thus, the user of a machine grouping algorithm has to use a layout algorithm such as CRAFT to develop a layout for the machines and cells. Moreover, material flow volume (which is an important factor in developing layout of cells and machines) is not considered in these algorithms.

In this chapter, we provide a brief discussion of the interactive decision support system (IDSS) and a detailed discussion of its interactive and menu-driven features. Specific details concerning the models and algorithms employed, performance comparison of the IDSS with other existing systems, a numerical example etc., are provided in Heragu and Kakuturi (1992). The IDSS, developed by Heragu and Kakuturi (1992),

		Machines					
		A	B	C	D	E	F
P							
a	1	1	0	2	0	3	0
r	2	0	1	0	2	0	0
t	3	0	0	1	0	2	0
s	4	1	0	0	0	3	2

Fig. 9.1 Routing of parts in a manufacturing system.

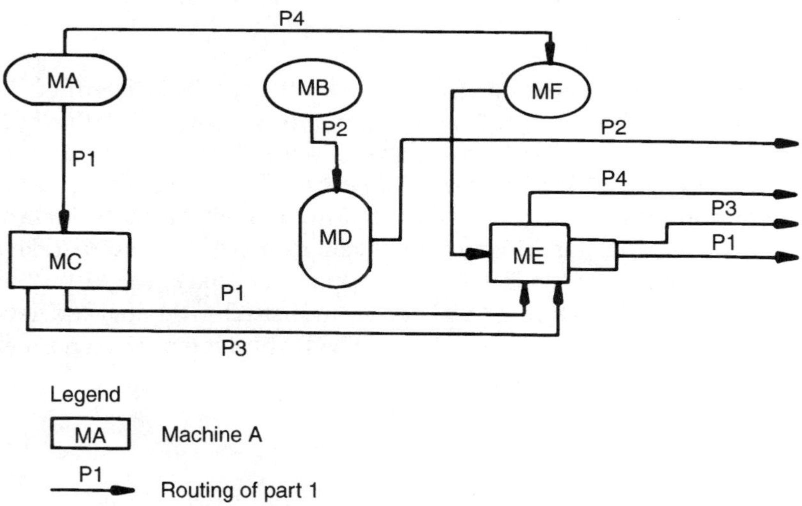

Fig. 9.2 Arrangement of machines in a job shop.

provides a single platform for performing machine grouping and layout.

The approach used by the IDSS follows three stages (flow chart, Fig. 9.4). In the first stage, grouping of machines into cells and the corresponding parts into part families is done. To do so, an efficient heuristic algorithm which considers many practical constraints is used. The IDSS considers alternative process plans and part routing information while grouping machines; thus, unlike past research, a non-binary part–machine processing indicator matrix is developed and used. Alternative process plans enable us to increase the throughput rate and decrease in-process inventory. In the second stage, the layout of machines within the

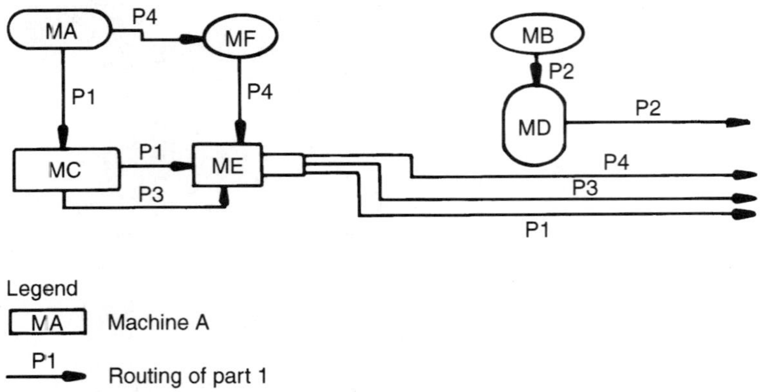

Fig. 9.3 Arrangement of machines in a cellular manufacturing system.

cells is determined. In the third stage, the layout of cells is determined. To develop a layout in stages 2 and 3, a modified version of the hybrid simulated annealing (HSA) algorithm discussed in Heragu and Alfa (1990) is used. This simulated-annealing-based algorithm has produced the best solutions compared to other algorithms available in the literature, for large test problems in which the facilities are of equal area (Heragu and Alfa, 1992). The IDSS has been applied satisfactorily to problems available in the literature as well as an industrial grouping and layout problem (Heragu and Kakuturi, 1992). The IDSS is interactive and menu-driven. It is explained in section 9.4.

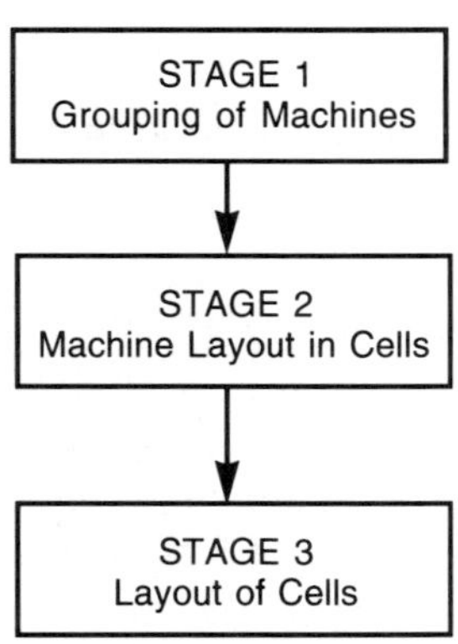

Fig. 9.4 Flow chart for three-stage approach.

9.2 LITERATURE REVIEW

Previous researchers have developed different algorithms to solve the machine grouping/part family formation problem. In this chapter, we provide a brief review of some popular algorithms. More detailed reviews can be obtained from other sources, for example, Heragu (1994). The algorithms developed can be broadly classified as production flow analysis, similarity coefficient, clustering analysis, graph theoretic, mathematical programming and networking methods.

Burbidge (1963, 1977) proposed a manual method in which production flow is analyzed by examining component route cards. This method leads to the identification of machine groups and corresponding part families. De Beer, Van Gerwen and De Witte (1976) and De Beer and De Witte (1978) used non-algorithmic machine grouping techniques based on visual examination. Obviously, such techniques are not feasible for large problems.

McAuley (1972) used similarity coefficients to form clusters. The similarity coefficient for a pair of machines is the number of parts visiting both machines divided by the number of parts visiting at least one machine. De Witte (1980) used similarity coefficients to describe the interrelations between operations. This information was used to form cell structures. Seifoddini and Wolfe (1986) also used similarity coefficients to identify machine groups and used threshold values to determine whether or not a machine is to be admitted to an existing group. This was done to overcome the drawback of similarity-coefficient-based methods. They also used efficient data storage and analysis techniques to improve the algorithm's performance.

McCormick, Schweitzer and White (1972) developed a general cluster analysis heuristic technique that maximizes the sum of the bond energies for each part–machine pair in the part–machine processing indicator matrix $[a_{ij}]$. The bond-energy for a pair i,j is given by $a_{ij}[a_{i,j+1} + a_{,j-1} + a_{i+1j} + a_{i-1j}]$.

Kusiak and Chow (1987) proposed a simple cluster identification algorithm for machine grouping. This algorithm restructures the part–machine processing indicator matrix efficiently by examining rows and columns alternately. However, the same algorithm was presented earlier by Iri (1968). Chan and Milner (1982) developed a direct clustering algorithm which also progressively restructures the part–machine processing indicator matrix. Chandrasekharan and Rajagopalan (1986) developed an algorithm called the ideal seed non-hierarchical clustering algorithm. They presented a graph theoretic formulation of the clustering problem and developed expressions for the upper limit on the number of groups. They used this limit and the clustering algorithm to identify part families and machine cells simultaneously. This method was extended, improved and presented as a zero–one data ideal seed algorithm for clustering (ZODIAC) in Chandrasekharan and Rajagopalan (1987). Srinivasan and Narendran (1991) developed a non-hierarchical clustering algorithm called GRAPHICS to solve the grouping problem. Their objective was to minimize exceptional elements and (indirectly) minimize machine idling. Initial seeds obtained from the assignment algorithm in Srinivasan *et al.* (1990) were used in the GRAPHICS algorithm. King (1980) proposed a rank order clustering (ROC) algorithm. The algorithm determines a 'binary value' for each row and column and iteratively rearranges rows and columns in descending order of their binary values to identify clusters. King (1980) also discussed other matrix restructuring methods and strategies to overcome clustering problems when the part–machine matrix has exceptional elements. King and Nakornchoi (1982) surveyed four different categories of methods – similarity coefficient methods, set-theoretic methods, evaluative methods and analytical methods. They also revised the ROC to reorder the part–machine processing indicator matrix based on the machine usage.

Boe and Cheng (1991) presented a close neighbor algorithm which rearranges a closeness matrix. This matrix shows, for each pair of machines, the corresponding closeness (which is a function of the parts processed by the pair). Using this matrix and a heuristic procedure, they identified clusters. The close neighbor algorithm was found to be superior with respect to solution quality and computation time for test problems available in the literature.

Rajagopalan and Batra (1975) used a graphical method to solve the machine grouping problem. Beginning with a graph with no arcs, their method adds one arc at a time, provided the similarity coefficient between the vertices that the arc connects is greater than a threshold value. Each vertex represents a machine and an arc connecting two vertices represents inclusion of the two machines in the same cell. Their method considers additional factors, such as upper limit on machine cell size and machine utilization in each cell. Askin and Chiu (1990) formulated a mathematical cost model for the machine grouping problem and solved it (suboptimally) using a two-phase graph partitioning algorithm. The first phase assigns parts to machines and the next phase assigns machines to cells.

Rajamani, Singh and Aneja (1992) developed a mathematical model for machine cell formation which considers setup times and costs. Srinivasan, Narendran and Mahadevan (1990) used an assignment model for solving the machine grouping problem. A similarity coefficient matrix (between the machines) is used in solving the assignment problem. Subtours obtained after solving the assignment problem are used as the basis for machine grouping.

Kaparthi and Suresh (1992) proposed a neural network approach to solve the machine-component cell formation. Binary inputs and the Carpenter–Grossberg network (Carpenter and Grossberg, 1988) were used in solving the problem. The Carpenter–Grossberg network was used as it can be trained without supervision. A parameter called the vigilance parameter was used in the network to control the size and formation of clusters.

Vohra *et al.* (1990) presented a network approach to represent a part–machine processing indicator matrix and partitioned it using a modified Gomory–Hu algorithm. Teng and Black (1990) developed a Petri-net model for a cellular manufacturing system to capture the dynamic behavior of the cells.

Ballakur and Steudel (1987) proposed a heuristic for part family/machine group formation. This technique adds machines to a cell based on cell work load and cell size restrictions. It is flexible and allows the user to control the admission of a machine to a cell. Logendran (1990) attempted to solve the machine grouping problem by considering the utilization of machines and minimization of intercell moves. Machines that are dissimilar with respect to the parts they process are selected and placed one in each cell. The remaining machines are added to these cells to minimize the total movement of parts. Parts are then assigned to appropriate cells. Vakharia and Wemmerlov (1990) presented a methodology to identify cells using an index based on operation sequences and material flow. They also suggested elimination of parts that visit the same machine more than once in their routing sequence.

In this section we have provided the reader with an overview of the various approaches to the machine grouping and part family formation problem. There are a number of survey papers available. Heragu (1994) also provides a list of other survey papers.

9.3 SOLUTION METHOD

Most researchers who have addressed the group technology problem have not fully recognized the use of part routings in the machine grouping problem. They have primarily used the binary part-machine processing indicator matrix. The IDSS presented in this chapter captures the part routings in the part–machine processing indicator matrix. Moreover, alternative process plans are explicitly considered while grouping machines. To our knowledge, the only other approach that considers alternative process plans is presented in Nagi, Harhalakis and Proth (1990). Although they use a non-binary matrix and consider multiple part routings to form machine cells, only the best routing (from the alternatives available) that minimizes the bottleneck machines and the exceptional parts is retained. Moreover, they use dummy parts to represent an alternative part routing which increases the size of the problem by the number of alternative routings. However, we consider all the alternative routings while grouping machines and parts. Our approach assumes that a part can use an alternative routing, if a required machine in the primary routing is busy or down for repair. Note that alternative routings enable us to reduce the waiting time of the part, increase productivity and also decrease in-process inventory. In the following discussion we list the data required by the IDSS. A sample data input file is shown in the appendix.

Part-routing information matrix

This matrix indicates the machines required to perform operations on each part as well as the sequence of these operations. Researchers have typically used a binary matrix to indicate whether or not a part visits a machine for an operation. Such a matrix does not capture the sequence of operations which is essential in determining interaction between machines. Information concerning interaction between machines is of course, used in determining layouts of machines and cells. The data representation and approach presented in this chapter is different. Figure 9.5 shows an example of the part-routing information matrix used. All alternative part routings are represented in the same row. The non-binary matrix has two additional columns – one for the quantity of each part demanded in a planning period (year or month) and the other for the batch size of each part. The batch size information is used to determine the number of trips required by the material handling carrier to transport

	Machine Type																
Part Type	1	2	3	4	5	6	7	8	9	10	11	12	4	7	11	Part Demand	Batch Size
1	1	0	0	2	0	0	0	3	4	4	0	0	0	0	0	2	1
2	1	0	0	2	0	0	3	5	0	0	0	0	4	6	0	3	1
3	1	2	0	3	0	0	4	5	6	0	0	0	0	0	0	1	1
4	1	0	0	2	0	0	3	3	4	0	0	0	0	0	0	3	1
5	1	0	0	0	0	2	4	0	5	3	0	0	0	0	0	2	1
6	0	0	0	0	0	1	3	4	5	2	0	0	0	0	0	1	1
7	0	0	0	2	0	1	0	3	4	0	0	0	0	0	0	2	1
8	0	3	1	5	2	4	0	6	7	0	0	0	0	0	0	1	1
9	0	0	1	4	2	3	0	5	6	0	0	0	0	0	0	1	1
10	1	0	0	1	0	0	2	4	0	0	0	0	3	0	0	2	1
11	0	0	0	0	0	1	0	0	0	0	0	0	0	0	0	3	1
12	0	0	0	0	0	0	2	0	0	0	1	3	0	0	0	1	1
13	0	0	0	0	0	0	0	0	0	0	1	2	0	0	0	1	1
14	0	0	0	0	0	0	2	0	0	3	1	0	0	0	0	3	1
15	1	0	0	0	0	0	2	0	0	4	3	6	0	0	5	1	1
16	1	0	0	0	0	0	2	0	0	4	3	6	0	0	5	2	1
17	0	0	0	0	0	0	2	0	0	0	1	3	0	0	0	1	1
18	0	0	0	0	0	1	2	0	0	3	0	0	0	0	0	3	1
19	0	0	0	0	0	0	0	0	0	0	0	1	0	0	0	2	1

Fig. 9.5 Part routing information matrix.

each part. This information is useful in calculating the flow of material between machines. The integer k in cell (i, j) of the matrix indicates that machine j is required for k^{th} operation on part i. The same integer k in two or more columns of a given row indicates the availability of an alternative process plan for that part. For example, a '1' in the first row and first column of the matrix indicates that part 1 is processed first on machine 1. A '2' in the first row and fourth column indicates that machine 4 is the second in the routing sequence of part 1. A '4' in the ninth and tenth columns of the first row indicate that part 1 has an alternative routing. Thus, the routings of part 1 are 1–4–8–9 and 1–4–8–10. If a part has two or more non-consecutive operations to be performed on the same machine, then an additional column corresponding to the dummy machine has to be created with the original machine number at the top of the column and the part-routing information included appropriately. For example, the routing of part 2 is 1–4–7–4–8–7. This is represented in the matrix by adding two columns corresponding to machine numbers 4 and 7. The dummy machines will be used only in the calculation of flow between the machines and will not increase the size of the problem. In general, if a machine is visited by a part for two or more non-consecutive operations, it is better to modify the process plan so that only consecutive operations are done on any machine; this decreases set-up time and increases machine utilization. The last two columns of the matrix indicate the demand for each part in the planning period and the batch size of each part carried by the material-handling equipment.

Machine relationship matrix

This matrix indicates whether there is any A or X-relationship between each machine pair. Figure 9.6 shows a sample machine relationship

Machine Type	1	2	3	4	5	6	7	8	9	10	11	12	4	7	11	12
1	0	0	0	A	0	0	0	0	0	0	0	0	0	0	0	0
2	0	0	0	0	0	0	0	0	0	0	0	0	0	0	0	0
3	0	0	0	0	0	0	0	0	0	0	0	0	0	0	0	0
4	A	0	0	0	X	0	0	0	0	0	0	0	0	0	0	0
5	0	0	0	X	0	0	0	0	0	0	0	0	0	0	0	0
6	0	0	0	0	0	0	0	0	0	0	0	0	0	0	0	0
7	0	0	0	0	0	0	0	0	0	0	0	0	0	0	0	0
8	0	0	0	0	0	0	0	0	0	0	0	0	0	0	0	0
9	0	0	0	0	0	0	0	0	0	0	0	0	0	0	0	0
10	0	0	0	0	0	0	0	0	0	0	0	0	0	0	0	0
11	0	0	0	0	0	0	0	0	0	0	0	0	0	0	0	0
12	0	0	0	0	0	0	0	0	0	0	0	0	0	0	0	0

Fig. 9.6 Machine relationship matrix.

matrix. *A* and *X* correspond to the adjacency relationships specified in Muther (1973), i.e. an *A* relationship between two machines indicates that the machines should be placed close to one another and an *X* indicates that they should be far apart. A 'O' in the matrix indicates that the placement of machines is to be determined by the layout algorithm.

Material handling cost matrix
This shows the cost of handling a unit load of material through a unit distance between machines *i* and *j*.

Alternative machine probability matrix
This matrix is used to indicate the probability that a part uses a machine in an alternative process plan (Fig. 9.7). The first row of the matrix in this figure indicates that part 1 has two routings, 1–4–8–9 and 1–4–8–10. '0.2' in the tenth column indicates that 20% of the time part 1 will be routed through machine 10 instead of machine 9 (because machine 9 is expected to be busy or down for repair).

Machine Type

Part Type	1	2	3	4	5	6	7	8	9	10	11	12	4	7	11
1	1	0	0	1	0	0	0	1	.8	.2	0	0	0	0	0
2	1	0	0	1	0	0	1	1	0	0	0	0	1	1	0
3	1	1	0	1	0	0	1	1	1	0	0	0	0	0	0
4	1	0	0	1	0	0	.9	.1	1	0	0	0	0	0	0
5	1	0	0	0	0	1	1	0	1	1	0	0	0	0	0
6	0	0	0	0	0	1	1	1	1	1	0	0	0	0	0
7	0	0	0	1	0	1	0	1	1	0	0	0	0	0	0
8	0	1	1	1	1	1	0	1	1	0	0	0	0	0	0
9	0	0	1	1	1	1	0	1	1	0	0	0	0	0	0
10	.05	0	0	.95	0	0	1	1	0	0	0	0	1	0	0
11	0	0	0	0	0	1	0	0	0	0	0	0	0	0	0
12	0	0	0	0	0	0	1	0	0	0	1	1	0	0	0
13	0	0	0	0	0	0	0	0	0	0	1	1	0	0	0
14	0	0	0	0	0	0	1	0	0	1	1	0	0	0	0
15	1	0	0	0	0	0	1	0	0	1	1	1	0	0	1
16	1	0	0	0	0	0	1	0	0	1	1	1	0	0	1
17	0	0	0	0	0	0	1	0	0	0	1	1	0	0	0
18	0	0	0	0	0	1	1	0	0	1	0	0	0	0	0
19	0	0	0	0	0	0	0	0	0	0	0	1	0	0	0

Fig. 9.7 Alternative machine probability matrix.

In addition to the above, the maximum number of machines allowed in a cell, maximum number of cells and the machine dimensions (i.e. length and width) must be provided.

The flow matrix which is used by the machine grouping and layout algorithms in IDSS is calculated from the part routing information matrix and the alternative machine probability matrix using the following notation and formula.

- f_{ij} is the number of trips made by the material handling equipment between machines i and j.
- m is the number of parts.
- D_k is the demand of part k in a given period.
- P_{ki} is the probability that part k is processed on machine i.
- B_k is the batch size of part k.
- R_{ki} is the operation number for the operation done on part k using machine i.

$$f_{ij} = \sum_{k=1}^{m} F_{ij}^{k} \tag{9.1}$$

$$F_{ij}^{k} = \begin{cases} \left(\frac{D_k p_{ki} p_{kj}}{B_k}\right) & \text{if } |R_{ki} - R_{kj}| = 1 \\ 0 & \text{otherwise} \end{cases} \tag{9.2}$$

Note that F_{ij}^{k} is defined only for consecutive operations. An input file for a sample problem is shown in the appendix. As illustrated in Fig. 9.4, the problem-solving approach employed by the IDSS follows the three stages explained below.

9.3.1 Stage 1 (grouping of machines into cells)

Stage 1 is concerned with the grouping of machines into cells and parts into families. The machine grouping algorithm used by the IDSS is very flexible and considers many important and practical design constraints. It is a modified and improved version of the algorithm developed by Heragu and Gupta (1994).

The design constraints considered are as follows.

1. Safety and technological requirements must be met.
2. The size of the cell cannot exceed a user-specified value.
3. The number of cells cannot exceed a user-specified value.

In addition to the above constraints, some additional features are incorporated. These include:

1. identification of bottleneck machines, which when duplicated in appropriate cells will minimize or eliminate intercellular movement of parts;
2. identification of exceptional (bottleneck) parts, which may be subcontracted or processed in a job shop;

3. consideration of material flow at every appropriate step to achieve the objective of minimizing inter- and intracellular material handling trips;
4. consideration of alternative process plans, which allows us to achieve a higher throughput rate. For example, in providing input data, we allow a part to be rerouted via another set of machines specified in a secondary process plan, if a machine in the primary process plan is busy or down for repair.

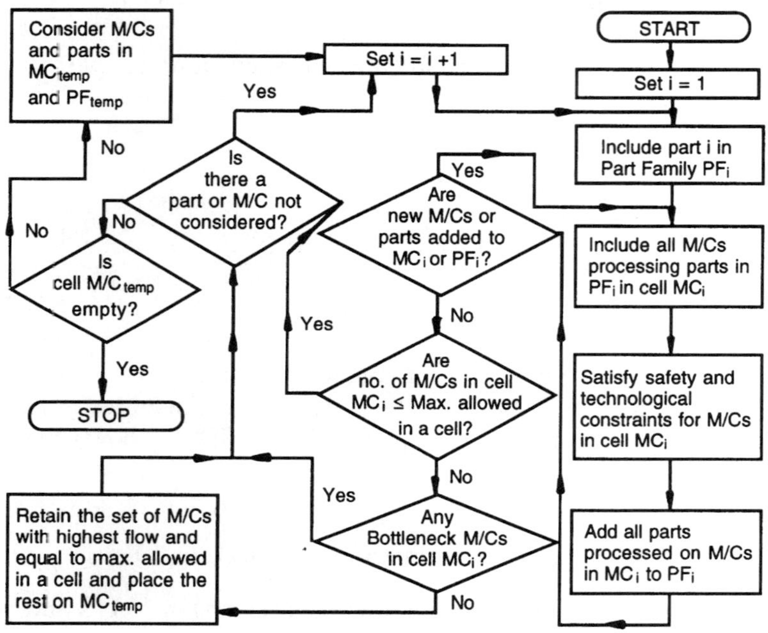

Fig. 9.8 Flow chart illustrating the machine grouping algorithm.

The flow chart for the above algorithm is shown in Fig. 9.8. Specific details concerning each step is provided in Heragu and Kakuturi (1992). In this section, only a brief explanation of the heuristic is provided. The algorithm begins by placing part 1 in part family PF_1. All the machines required to process parts in PF_1, including the machines in the alternative process plan of the part are placed in cell MC_1. Then, the heuristic attempts to satisfy the safety and technological constraints specified in the relationship matrix. The parts processed on machines in MC_1 are added to PF_1. Adding parts to PF_1 and machines to MC_1 is repeated until no new parts or machines can be added. If the number of machines in the cell MC_1 satisfy the cell size constraint, then the machine not considered thus far by the grouping algorithm is selected and placed in the next cell; corresponding parts are added to the next part family. The above steps are repeated for this cell and part family. If the cell size constraint is not satisfied for MC_1, then the algorithm tries to find bottleneck machines or exceptional parts to partition the cell into two or more cells of smaller size. If this is not possible, then the machines are partitioned on the basis of flow considerations. To do so, the heuristic algorithm identifies two machines that have the highest flow between them and places them in the current cell; all others are removed and placed in a temporary cell MC_{check}. Next, machines are added one at a time to the current cell such that the machine added has the greatest interaction (flow) with machines already in the current cell. This is done until the number of machines in the current cell is equal to the maximum allowed.

Note that the above algorithm is a heuristic attempt at partitioning a larger cell (with say n machines) into two or more cells of the required size (less than or equal to r machines) and is based on material flow consideration. The user has the option of optimally choosing the combination of r machines (amongst the n machines available) that have the greatest interaction between them. This involves checking ${}^{n}C_{r}$ combinations which obviously requires much CPU time, especially if n is large. Under such circumstances, it is better to use the heuristic.

In a similar manner, other cells and part families are created by considering the remaining machines and parts. After all machines have been considered by the grouping algorithm, the above procedure is repeated for only the machines in the temporary cell and the corresponding parts. After all the cells are determined, the constraint on the number of cells is satisfied (if necessary) by combining the cells with most interaction between them. If this constraint violates the cell size constraint, then obviously one of them has to be relaxed and the user is notified accordingly. If necessary, an additional post-processing step that allocates available extra copies of machines to cells so as to minimize intercellular movement of parts may be included.

Although the above algorithm is based on the one presented in Heragu and Gupta (1994), there are a number of significant improvements. First, the data representation is different from that used by Heragu and Gupta (1994). The latter heuristic uses a binary part–machine processing indicator matrix. Although Heragu and Gupta (1994) indirectly used the part-routing information to evaluate the flow matrix, the importance of part routings has not been fully exploited. The other notable improvements presented in this chapter include:

1. the consideration of alternative process plans which is a useful practical issue;
2. modification of a specific step to decrease intercellular flow;

3. attempts to determine two bottleneck machines simultaneously;
4. attempts to determine exceptional parts;
5. addition of an efficient heuristic to identify combinations of *r* machines amongst *n* available ones.

The machine cells identified by the IDSS for test problem 2 in Heragu and Kakuturi (1992) are shown in Fig. 9.9. Although there are a number of parts visiting multiple cells for this problem, the solution is such that the material handling carrier makes very few trips between the three cells.

Machine Cell	1–1, 4, 7, 8, 9, 10 and 11
Machine Cell	2–2, 3, 5 and 6
Machine Cell	3–12

Fig. 9.9 List of machines in three cells.

9.3.2 Stage 2 (machine layout in cells)

Once the grouping of machines into cells is completed, the layout of machines within the cells has to be determined. A number of algorithms for solving the layout problem are available for this purpose (see Heragu, 1992, for a brief review). As mentioned previously, CRAFT could be used. To efficiently use the data generated by the machine grouping algorithm and efficiently solve the layout problem, the IDSS uses the hybrid simulated annealing (HSA) algorithm presented in Heragu and Alfa (1992). The HSA algorithm generates an initial solution using a modified penalty (MP) algorithm and improves it using two improvement algorithms. The improvement algorithms are two-way greedy exchange and simulated annealing (SA) algorithms. The HSA algorithm has produced better results compared to other algorithms for equal area test problems (Heragu and Alfa, 1992). The two-way greedy exchange algorithm examines the pairwise exchange between positions of machines. Any starting layout can be taken as the initial solution to this algorithm. If the exchange leads to a solution with a lower objective function value (OFV), the exchange is made; otherwise the next pair of machines is considered. Since all the machines may not have same shape or area, the exchange of the positions of two machines requires a modification step described later in this section to attain feasibility of layout, i.e. to avoid overlap of two or more machines. The final solution obtained from this algorithm is provided as input to the SA algorithm for further improvement. More details on the HSA, parameter settings for the SA algorithm and layout modification step necessary when using the two-way and SA algorithms can be found in Heragu and Kakuturi (1992).

The final layout of each cell obtained from this algorithm is appropriately scaled and printed. The OFV of the layout, dimensions of the cell and clearance between the machines within the cell are also calculated and printed. The user has the option to zoom the layout. Moreover, if the layout determined by the HSA algorithm does not meet the facility designer's requirements (i.e. building dimensions exceeded, space not fully utilized, machines requiring the same operator not placed adjacently, etc.), it can be changed by varying the maximum number of machines allowed in any row and/or column of the cell. Thus the designer can try different combinations and retain the one that best satisfies his/her requirements. For the three machine groups in Fig. 9.9, the machine layout within each cell generated by the IDSS is shown in Fig. 9.10.

```
      Cell 1                   Cell 2              Cell 3

4 4 4 4 7 7 8 8 9 9        6 6 6 6 2 2             1212
4 4 4 4 7 7 8 8 9 9        6 6 6 6 2 2             1212
1 111111010                6 6 6 6 2 2             1212
1 111111010                6 6 6 6
1 1                        3 3 3 3 5 5
1 1                        3 3 3 3 5 5
                           3 3 3 3
                           3 3 3 3
```

Fig. 9.10 Layout of machines in three cells.

9.3.3 Stage 3 (layout of cells)

After the layout of machines within each cell is finalized in stage 2, the dimensions of each cell and the flow between the cells is calculated. The same algorithm discussed in stage 2 is used to determine the layout of cells. Since the number of cells is relatively small, a near-optimum layout of cells can be generated. Again, the final layout of the cells can be changed to satisfy the user requirements, by changing the maximum number of cells allowed in any row and/or column. Even after the final layout of cells is determined, the user has the flexibility of changing the

```
3 3 2 2 2 2 2 2
3 3 2 2 2 2 2 2
3 3 2 2 2 2 2 2
3 3 2 2 2 2 2 2
    2 2 2 2 2 2
1 1 1 1 1 1 1 1 1 1
1 1 1 1 1 1 1 1 1 1
1 1 1 1 1 1 1 1 1 1
1 1 1 1 1 1 1 1 1 1
1 1 1 1 1 1 1 1 1 1
1 1 1 1 1 1 1 1 1 1
```

Fig. 9.11 Layout of cells with respect to each other.

layout of machines within any cell. The final layout retained by the user is printed with appropriate scaling. The layout of cells is provided in Fig 9.11.

9.4 INTERACTIVE DECISION SUPPORT SYSTEM

In this section, we discuss the menu-driven features of IDSS. It is interactive and user friendly. The main features of the program are discussed below. At every stage, the program prompts the user to select from the menu screen options. The program starts with the main menu which has nine different options. The main menu screen is shown in Fig 9.12. Options 1 to 5 are used for data manipulation or correction, 6 to 8 to run the grouping and layout algorithms and 9 to quit the program.

The main menu options 1, 2, 3 and 5 have submenus shown in Figures 9.13–9.16 respectively. The machine menu can be used to add or delete a machine. The program allows the user to name a machine with a six-character (or less) alphanumeric code of his/her choice. It automatically checks the database for the existence of this machine and prompts the user for appropriate data. It then updates the complete database. To delete a machine, the program checks the database for any parts manufactured using the machine to be deleted. If so, it requires the user to modify the corresponding part routings (so that the modified routings do not include the machine to be deleted) before deleting the machine from the database.

The part menu can be used to add or delete a part or to change the part routings. The program accepts ten alphanumeric character names for the part after verifying that it does not exist in the database. It prompts the user for part relevant information and updates the database. All the new machines used by the part which do not exist in the database have to be added before adding the part. If not, the program suggests to, and

```
     MAIN MENU
1.   MACHINE (Add OR Delete)
2.   PART (Add OR Delete OR Update)
3.   CHANGE RELATIONSHIPS or MATERIAL HANDLING
     RATE BETWEEN MACHINES
4.   CHANGE MAXIMUM MACHINES ALLOWED IN A CELL
5.   CHANGE MACHINE DIMENSIONS (Orientations)
6.   RUN GROUPING PROGRAM
7.   RUN LAYOUT PROGRAM
8.   RUN BOTH (Grouping AND Layout)
```

Fig. 9.12 Main menu.

```
     MACHINE MENU
1.   ADD A MACHINE
2.   DELETE A MACHINE
3.   MAIN MENU
```

Fig. 9.13 Machine menu.

```
     PART MENU
1.   ADD A PART
2.   DELETE A PART
3.   CHANGE A PART ROUTING
4.   MAIN MENU
```

Fig. 9.14 Part menu.

requires) the user to do so. The relationship menu can be used to change the relationship or material handling rate between the machines. The machine dimensions menu can be used to change the dimensions of the machine or to change the orientation of the machine (i.e., to swap the length and width of the machine).

The database is updated after every data correction step and the new data is used to run the algorithms. The machine grouping algorithm and the machine layout algorithm can be run sequentially as illustrated in Fig 9.4, or independently. The results obtained are written to different files, and the corresponding names are printed on the computer screen after the algorithms are run. The final updated data is written to a file different

```
      RELATIONSHIP MENU
1.    CHANGE RELATIONSHIPS
2.    CHANGE MATERIAL HANDLING RATES
3.    MAIN MENU
```

Fig. 9.15 Relationship menu.

```
      MACHINE DIMENSIONS MENU
1.    CHANGE MACHINE DIMENSIONS (Length OR Width)
2.    CHANGE ORIENTATIONS or SWAP LENGTH AND WIDTH
3.    MAIN MENU
```

Fig. 9.16 Machine dimension menu.

from the initial data file. The new data file can be used as input to the program, the next time. The menu-driven and interactive features of the program make it easy to use in real-world problems.

9.5 CONCLUSIONS

Traditionally, the machine grouping and layout problems have been addressed independently. Since grouping decisions have a significant impact on layout decisions, it appears that there is much to gain by addressing the two issues simultaneously. In this chapter, we discuss a three-stage approach that attempts to integrate the grouping and layout problems. Our objective is not only to identify machine cells and corresponding part families but also to determine a near-optimal layout of machines within each cell and the cells themselves. Although the grouping and layout problems are solved in three stages, material flow considerations are incorporated while determining machine groups. To our knowledge, this is the first attempt in which the two issues have been addressed. Most grouping algorithms developed so far neither consider material flow volume nor provide a layout of the cells. Hence, another layout algorithm such as CRAFT has to be used to determine the layout. The three-stage approach discussed in this chapter has been used to solve a number of test problems and an industrial problem. It has produced satisfactory results for all these problems. The important features of this approach are the flexibility to change the design constraints, sensitivity analysis and consideration of alternative process plans in machine grouping. The additional advantage is the interactive program which is easy to use.

Apart from the above advantages, there are also some disadvantages of the algorithm. The layout determined by the algorithm may have unused (empty) spaces because of the machine dimensions. Empty space can be minimized but the user must try different combinations of the maximum number of machines in a row and/or column of a cell in order to do so. The user can try changing the orientations of machines interactively to minimize the empty spaces. This may also result in a solution with lower objective function value.

ACKNOWLEDGEMENTS

Research presented in this chapter was supported in part by grants from the New York State Science and Technology Foundation and Welch Allyn Medical Division, NY, USA. This chapter is based on the authors technical report, 'Grouping and Placement of Machine Cells', DSES Technical Report 37–92–341, Rensselaer Polytechnic Institute, Troy, NY, USA.

APPENDIX

Input data file for a sample problem

```
NMACHINES
27
NPARTS
40
NDUM
0
PART TO MACHINE USE MATRIX
```

	Machines																												
	1	2	3	4	5	6	7	8	9	10	11	12	13	14	15	16	17	18	19	20	21	22	23	24	25	26	27	Demand	Batch
	0	0	0	0	0	3	3	0	2	0	0	1	0	0	0	0	0	0	0	0	4	0	0	0	0	0	0	2	1
	3	0	0	0	0	1	1	0	0	0	0	2	0	0	0	0	0	0	0	0	0	0	0	4	0	0	0	2	1
	1	0	0	0	0	0	4	0	2	0	0	3	0	0	0	0	0	0	0	0	0	0	5	5	0	0	0	1	1
	1	0	0	0	0	3	3	0	0	0	0	2	0	0	0	0	0	0	0	0	4	0	0	0	0	0	0	2	1
	2	0	0	0	0	0	0	0	1	0	0	3	0	0	0	0	0	4	4	0	0	0	0	5	0	0	0	2	1
	0	3	0	0	2	1	1	0	0	0	0	0	0	0	0	0	0	0	0	0	0	0	0	0	4	0	0	3	1
	0	0	0	0	1	4	4	0	0	0	0	0	0	0	0	2	0	0	3	3	0	0	0	0	5	0	0	2	1
	0	3	0	0	0	2	2	0	0	0	0	0	0	0	0	1	0	0	0	0	0	0	4	0	0	0	0	2	1
	0	2	0	0	0	4	4	0	0	0	0	0	0	0	0	1	0	0	3	3	0	0	5	0	0	0	0	1	1
	0	1	0	0	3	4	0	0	0	0	0	0	0	0	0	2	0	0	0	0	0	0	0	0	5	6	0	3	1
	0	0	2	0	0	0	0	1	0	0	3	0	0	0	0	0	0	4	4	0	0	0	0	0	0	5	0	3	1
	0	0	1	0	0	0	0	2	0	0	0	0	0	0	0	0	0	3	3	0	0	0	0	0	5	0	4	2	1
P	0	0	3	0	0	0	0	2	0	0	1	0	0	0	0	0	0	4	4	0	0	0	0	0	0	0	0	4	1
a	0	0	0	0	0	0	0	0	0	1	0	0	0	3	0	0	4	0	2	2	0	5	0	0	0	0	6	6	2
r	0	0	0	0	0	0	0	0	0	3	0	0	0	0	0	0	2	1	1	0	0	0	0	0	0	0	4	2	2
t	0	0	0	0	0	0	0	0	0	2	0	0	0	3	0	0	0	1	1	0	0	0	4	0	0	0	0	9	3
s	0	0	0	0	0	0	0	0	0	3	0	0	0	2	0	0	1	0	0	0	0	4	0	0	0	0	4	2	1
	0	0	0	2	0	0	0	0	0	0	0	0	1	0	3	0	0	0	0	0	0	0	0	0	4	0	0	1	1
	0	0	0	1	0	0	0	0	0	0	0	0	4	0	3	0	0	0	0	0	0	0	0	0	2	0	0	8	2
	0	0	0	2	0	0	0	0	0	0	0	0	0	0	1	0	0	0	0	0	0	0	0	0	3	0	0	3	1
	0	0	0	0	0	0	3	0	2	0	0	1	0	0	0	0	0	0	0	0	4	0	0	0	0	0	0	2	1
	3	0	0	0	0	4	4	0	0	0	0	2	0	0	0	0	0	0	0	0	0	0	0	1	0	0	0	2	1
	3	0	0	0	0	0	5	0	2	0	0	1	0	0	0	0	0	0	0	0	0	0	4	4	0	0	0	3	1
	1	0	0	0	0	0	0	0	0	0	0	2	0	0	0	0	0	0	0	0	3	0	0	0	0	0	0	4	2
	1	0	0	0	0	0	0	0	2	0	0	3	0	0	0	0	0	0	4	0	0	0	0	5	0	0	0	8	4
	0	2	0	0	0	1	1	0	0	0	0	0	0	0	0	0	0	0	0	0	0	0	0	0	3	0	0	9	3
	0	0	0	0	1	2	2	0	0	0	0	0	0	0	0	3	0	0	0	4	0	0	0	0	5	0	0	5	1
	0	3	0	0	0	1	1	0	0	0	0	0	0	0	0	2	0	0	0	0	0	0	4	4	0	0	0	2	1
	0	1	0	0	0	4	4	0	0	0	0	0	0	0	0	5	0	0	3	3	0	0	2	0	0	0	0	3	1
	0	1	0	0	3	4	0	0	0	0	0	0	0	0	0	6	0	0	0	0	0	0	0	0	5	2	0	1	1
	0	0	4	0	0	0	0	2	0	0	3	0	0	0	0	0	0	1	0	0	0	0	0	0	5	5	0	2	1
	0	0	5	0	0	0	0	2	0	0	0	0	0	0	0	0	0	3	0	0	0	0	0	0	4	0	1	1	1
	0	0	3	0	0	0	0	2	0	0	1	0	0	0	0	0	0	4	4	0	0	0	0	0	0	0	0	4	2
	0	0	0	0	0	0	0	0	0	4	0	0	0	1	0	0	5	0	2	2	0	3	0	0	0	0	6	6	2

Input data file for a sample problem *contd*

```
0 0 0 0 0 0 0 0 0  3  0  0  0  0  0  0  4  1  1  0  0  0  0  0  0  0  2   1   1
0 0 0 0 0 0 0 0 0  3  0  0  0  2  0  0  0  0  1  0  0  0  4  0  0  0  0   9   3
0 0 0 0 0 0 0 0 0  2  0  0  0  1  0  0  3  0  0  0  0  4  0  0  0  0  4   6   3
0 0 0 0 0 0 0 0 0  0  0  0  1  0  3  0  0  0  0  0  0  0  0  0  2  0  0   2   1
0 0 0 1 0 0 0 0 0  0  0  0  4  0  3  0  0  0  0  0  0  0  0  0  2  0  0   3   1
0 0 0 2 0 0 0 0 0  0  0  0  4  0  1  0  0  0  0  0  0  0  0  0  3  0  0   4   4

ALTERNATE MACHINE PROBABILITY MATRIX
                    Machines
  0.0 0.0 0.0 0.0 0.0 0.1 0.9 0.0 1.0 0.0 0.0 1.0 0.0 0.0 0.0 0.0 0.0 0.0 0.0 0.0 1.0 0.0 0.0 0.0 0.0 0.0 0.0
  1.0 0.0 0.0 0.0 0.0 0.2 0.8 0.0 0.0 0.0 0.0 1.0 0.0 0.0 0.0 0.0 0.0 0.0 0.0 0.0 0.0 0.0 0.0 1.0 0.0 0.0 0.0
  1.0 0.0 0.0 0.0 0.0 0.0 1.0 0.0 1.0 0.0 0.0 1.0 0.0 0.0 0.0 0.0 0.0 0.0 0.0 0.0 0.0 0.0 0.1 0.9 0.0 0.0 0.0
  1.0 0.0 0.0 0.0 0.0 0.3 0.7 0.0 0.0 0.0 0.0 1.0 0.0 0.0 0.0 0.0 0.0 0.0 0.0 0.0 1.0 0.0 0.0 0.0 0.0 0.0 0.0
  1.0 0.0 0.0 0.0 0.0 0.0 0.0 0.0 1.0 0.0 0.0 1.0 0.0 0.0 0.0 0.0 0.0 0.1 0.9 0.0 0.0 0.0 0.0 1.0 0.0 0.0 0.0
  0.0 1.0 0.0 0.0 1.0 0.8 0.2 0.0 0.0 0.0 0.0 0.0 0.0 0.0 0.0 0.0 0.0 0.0 0.0 0.0 0.0 0.0 0.0 0.0 1.0 0.0 0.0
  0.0 0.0 0.0 0.0 1.0 0.9 0.1 0.0 0.0 0.0 0.0 0.0 0.0 0.0 0.0 1.0 0.0 0.0 0.1 0.9 0.0 0.0 0.0 0.0 1.0 0.0 0.0
  0.0 1.0 0.0 0.0 0.0 0.9 0.1 0.0 0.0 0.0 0.0 0.0 0.0 0.0 0.0 1.0 0.0 0.0 0.0 0.0 0.0 0.0 1.0 0.0 0.0 0.0 0.0
P 0.0 1.0 0.0 0.0 0.0 0.8 0.2 0.0 0.0 0.0 0.0 0.0 0.0 0.0 0.0 1.0 0.0 0.0 0.2 0.8 0.0 0.0 1.0 0.0 0.0 0.0 0.0
a 0.0 1.0 0.0 0.0 1.0 1.0 0.0 0.0 0.0 0.0 0.0 0.0 0.0 0.0 0.0 1.0 0.0 0.0 0.0 0.0 0.0 0.0 0.0 0.0 1.0 1.0 0.0
r 0.0 0.0 1.0 0.0 0.0 0.0 0.0 1.0 0.0 0.0 1.0 0.0 0.0 0.0 0.0 0.0 0.0 0.9 0.1 0.0 0.0 0.0 0.0 0.0 0.0 1.0 0.0
t 0.0 0.0 1.0 0.0 0.0 0.0 0.0 1.0 0.0 0.0 0.0 0.0 0.0 0.0 0.0 0.0 0.0 0.8 0.2 0.0 0.0 0.0 0.0 0.0 1.0 0.0 1.0
s 0.0 0.0 1.0 0.0 0.0 0.0 0.0 1.0 0.0 0.0 1.0 0.0 0.0 0.0 0.0 0.0 0.0 0.9 0.1 0.0 0.0 0.0 0.0 0.0 0.0 0.0 0.0
  0.0 0.0 0.0 0.0 0.0 0.0 0.0 0.0 0.0 1.0 0.0 0.0 0.0 1.0 0.0 0.0 1.0 0.0 0.9 0.1 0.0 1.0 0.0 0.0 0.0 0.0 1.0
  0.0 0.0 0.0 0.0 0.0 0.0 0.0 0.0 0.0 1.0 0.0 0.0 0.0 0.0 0.0 0.0 1.0 0.2 0.8 0.0 0.0 0.0 0.0 0.0 0.0 0.0 1.0
  0.0 0.0 0.0 0.0 0.0 0.0 0.0 0.0 0.0 1.0 0.0 0.0 0.0 1.0 0.0 0.0 0.0 0.1 0.9 0.0 0.0 0.0 1.0 0.0 0.0 0.0 0.0
  0.0 0.0 0.0 0.0 0.0 0.0 0.0 0.0 0.0 1.0 0.0 0.0 0.0 1.0 0.0 0.0 1.0 0.0 0.0 0.0 0.0 0.2 0.0 0.0 0.0 0.0 0.8
  0.0 0.0 0.0 1.0 0.0 0.0 0.0 0.0 0.0 0.0 0.0 0.0 1.0 0.0 1.0 0.0 0.0 0.0 0.0 0.0 0.0 0.0 0.0 0.0 1.0 0.0 0.0
  0.0 0.0 0.0 1.0 0.0 0.0 0.0 0.0 0.0 0.0 0.0 0.0 1.0 0.0 1.0 0.0 0.0 0.0 0.0 0.0 0.0 0.0 0.0 0.0 1.0 0.0 0.0
  0.0 0.0 0.0 1.0 0.0 0.0 0.0 0.0 0.0 0.0 0.0 0.0 0.0 0.0 1.0 0.0 0.0 0.0 0.0 0.0 0.0 0.0 0.0 0.0 1.0 0.0 0.0
  0.0 0.0 0.0 0.0 0.0 0.0 1.0 0.0 1.0 0.0 0.0 1.0 0.0 0.0 0.0 0.0 0.0 0.0 0.0 0.0 1.0 0.0 0.0 0.0 0.0 0.0 0.0
  1.0 0.0 0.0 0.0 0.0 0.1 0.9 0.0 0.0 0.0 0.0 1.0 0.0 0.0 0.0 0.0 0.0 0.0 0.0 0.0 0.0 0.0 0.0 1.0 0.0 0.0 0.0
  1.0 0.0 0.0 0.0 0.0 0.0 1.0 0.0 1.0 0.0 0.0 1.0 0.0 0.0 0.0 0.0 0.0 0.0 0.0 0.0 0.0 0.0 0.1 0.9 0.0 0.0 0.0
  1.0 0.0 0.0 0.0 0.0 0.0 0.0 0.0 0.0 0.0 0.0 1.0 0.0 0.0 0.0 0.0 0.0 0.0 0.0 0.0 1.0 0.0 0.0 0.0 0.0 0.0 0.0
  1.0 0.0 0.0 0.0 0.0 0.0 0.0 0.0 1.0 0.0 0.0 1.0 0.0 0.0 0.0 0.0 0.0 0.0 1.0 0.0 0.0 0.0 0.0 1.0 0.0 0.0 0.0
  0.0 1.0 0.0 0.0 0.0 0.8 0.2 0.0 0.0 0.0 0.0 0.0 0.0 0.0 0.0 0.0 0.0 0.0 0.0 0.0 0.0 0.0 0.0 0.0 1.0 0.0 0.0
  0.0 0.0 0.0 0.0 1.0 0.8 0.2 0.0 0.0 0.0 0.0 0.0 0.0 0.0 0.0 1.0 0.0 0.0 0.0 1.0 0.0 0.0 0.0 0.0 1.0 0.0 0.0
  0.0 1.0 0.0 0.0 0.0 0.9 0.1 0.0 0.0 0.0 0.0 0.0 0.0 0.0 0.0 1.0 0.0 0.0 0.0 0.0 0.0 0.0 0.9 0.1 0.0 0.0 0.0
  0.0 1.0 0.0 0.0 0.0 0.8 0.2 0.0 0.0 0.0 0.0 0.0 0.0 0.0 0.0 1.0 0.0 0.0 0.2 0.8 0.0 0.0 1.0 0.0 0.0 0.0 0.0
  0.0 1.0 0.0 0.0 1.0 1.0 0.0 0.0 0.0 0.0 0.0 0.0 0.0 0.0 0.0 1.0 0.0 0.0 0.0 0.0 0.0 0.0 0.0 0.0 1.0 1.0 0.0
  0.0 0.0 1.0 0.0 0.0 0.0 0.0 1.0 0.0 0.0 1.0 0.0 0.0 0.0 0.0 0.0 0.0 1.0 0.0 0.0 0.0 0.0 0.0 0.0 0.0 0.8 0.2
  0.0 0.0 1.0 0.0 0.0 0.0 0.0 1.0 0.0 0.0 0.0 0.0 0.0 0.0 0.0 0.0 0.0 1.0 0.0 0.0 0.0 0.0 0.0 0.0 1.0 0.0 1.0
  0.0 0.0 1.0 0.0 0.0 0.0 0.0 1.0 0.0 0.0 1.0 0.0 0.0 0.0 0.0 0.0 0.0 0.9 0.1 0.0 0.0 0.0 0.0 0.0 0.0 0.0 0.0
```

Input data file for a sample problem *contd*

```
0.0 0.0 0.0 0.0 0.0 0.0 0.0 0.0 0.0 1.0 0.0 0.0 0.0 1.0 0.0 0.0 1.0 0.0 0.9 0.1 0.0 1.0 0.0 0.0 0.0 0.0 1.0
0.0 0.0 0.0 0.0 0.0 0.0 0.0 0.0 0.0 1.0 0.0 0.0 0.0 0.0 0.0 0.0 1.0 0.1 0.9 0.0 0.0 0.0 0.0 0.0 0.0 0.0 1.0
0.0 0.0 0.0 0.0 0.0 0.0 0.0 0.0 0.0 1.0 0.0 0.0 0.0 1.0 0.0 0.0 0.0 0.0 1.0 0.0 0.0 0.0 1.0 0.0 0.0 0.0 0.0
0.0 0.0 0.0 0.0 0.0 0.0 0.0 0.0 0.0 1.0 0.0 0.0 0.0 1.0 0.0 0.0 1.0 0.0 0.0 0.0 0.0 0.2 0.0 0.0 0.0 0.0 0.8
0.0 0.0 0.0 0.0 0.0 0.0 0.0 0.0 0.0 0.0 0.0 0.0 1.0 0.0 1.0 0.0 0.0 0.0 0.0 0.0 0.0 0.0 0.0 0.0 1.0 0.0 0.0
0.0 0.0 0.0 1.0 0.0 0.0 0.0 0.0 0.0 0.0 0.0 0.0 1.0 0.0 1.0 0.0 0.0 0.0 0.0 0.0 0.0 0.0 0.0 0.0 1.0 0.0 0.0
0.0 0.0 0.0 1.0 0.0 0.0 0.0 0.0 0.0 0.0 0.0 0.0 1.0 0.0 1.0 0.0 0.0 0.0 0.0 0.0 0.0 0.0 0.0 0.0 1.0 0.0 0.0
MATERIAL HANDLING RATE MATRIX
                   Machines
  0 1 1 1 1 1 1 1 1 1 1 1 1 1 1 1 1 1 1 1 1 1 1 1 1 1 1
  1 0 1 1 1 1 1 1 1 1 1 1 1 1 1 1 1 1 1 1 1 1 1 1 1 1 1
  1 1 0 1 1 1 1 1 1 1 1 1 1 1 1 1 1 1 1 1 1 1 1 1 1 1 1
M 1 1 1 0 1 1 1 1 1 1 1 1 1 1 1 1 1 1 1 1 1 1 1 1 1 1 1
a 1 1 1 1 0 1 1 1 1 1 1 1 1 1 1 1 1 1 1 1 1 1 1 1 1 1 1
c 1 1 1 1 1 0 1 1 1 1 1 1 1 1 1 1 1 1 1 1 1 1 1 1 1 1 1
h 1 1 1 1 1 1 0 1 1 1 1 1 1 1 1 1 1 1 1 1 1 1 1 1 1 1 1
i 1 1 1 1 1 1 1 0 1 1 1 1 1 1 1 1 1 1 1 1 1 1 1 1 1 1 1
n 1 1 1 1 1 1 1 1 0 1 1 1 1 1 1 1 1 1 1 1 1 1 1 1 1 1 1
e 1 1 1 1 1 1 1 1 1 0 1 1 1 1 1 1 1 1 1 1 1 1 1 1 1 1 1
s 1 1 1 1 1 1 1 1 1 1 0 1 1 1 1 1 1 1 1 1 1 1 1 1 1 1 1
  1 1 1 1 1 1 1 1 1 1 1 0 1 1 1 1 1 1 1 1 1 1 1 1 1 1 1
  1 1 1 1 1 1 1 1 1 1 1 1 0 1 1 1 1 1 1 1 1 1 1 1 1 1 1
  1 1 1 1 1 1 1 1 1 1 1 1 1 0 1 1 1 1 1 1 1 1 1 1 1 1 1
  1 1 1 1 1 1 1 1 1 1 1 1 1 1 0 1 1 1 1 1 1 1 1 1 1 1 1
  1 1 1 1 1 1 1 1 1 1 1 1 1 1 1 0 1 1 1 1 1 1 1 1 1 1 1
  1 1 1 1 1 1 1 1 1 1 1 1 1 1 1 1 0 1 1 1 1 1 1 1 1 1 1
  1 1 1 1 1 1 1 1 1 1 1 1 1 1 1 1 1 0 1 1 1 1 1 1 1 1 1
  1 1 1 1 1 1 1 1 1 1 1 1 1 1 1 1 1 1 0 1 1 1 1 1 1 1 1
  1 1 1 1 1 1 1 1 1 1 1 1 1 1 1 1 1 1 1 0 1 1 1 1 1 1 1
  1 1 1 1 1 1 1 1 1 1 1 1 1 1 1 1 1 1 1 1 0 1 1 1 1 1 1
  1 1 1 1 1 1 1 1 1 1 1 1 1 1 1 1 1 1 1 1 1 0 1 1 1 1 1
  1 1 1 1 1 1 1 1 1 1 1 1 1 1 1 1 1 1 1 1 1 1 0 1 1 1 1
  1 1 1 1 1 1 1 1 1 1 1 1 1 1 1 1 1 1 1 1 1 1 1 0 1 1 1
  1 1 1 1 1 1 1 1 1 1 1 1 1 1 1 1 1 1 1 1 1 1 1 1 0 1 1
  1 1 1 1 1 1 1 1 1 1 1 1 1 1 1 1 1 1 1 1 1 1 1 1 1 0 1
  1 1 1 1 1 1 1 1 1 1 1 1 1 1 1 1 1 1 1 1 1 1 1 1 1 1 0
RELATIONSHIP MATRIX
             Machines
  0 0 0 0 0 0 0 0 0 0 0 0 0 0 0 0 0 0 0 0 0 0 0 0 0 0 0
  0 0 0 0 0 0 0 0 0 0 0 0 0 0 0 0 0 0 0 0 0 0 0 0 0 0 0
  0 0 0 0 0 0 0 0 0 0 0 0 0 0 0 0 0 0 0 0 0 0 0 0 0 0 0
  0 0 0 0 0 0 0 0 0 0 0 0 0 0 0 0 0 0 0 0 0 0 0 0 0 0 0
  0 0 0 0 0 0 0 0 0 0 0 0 0 0 0 0 0 0 0 0 0 0 0 0 0 0 0
```

Input data file for a sample problem *contd*

```
  0 0 0 0 0 0 0 0 0 0 0 0 0 0 0 0 0 0 0 0 0 0 0 0 0 0 0
  0 0 0 0 0 0 0 0 0 0 0 0 0 0 0 0 0 0 0 0 0 0 0 0 0 0 0
M 0 0 0 0 0 0 0 0 0 0 0 0 0 0 0 0 0 0 0 0 0 0 0 0 0 0 0
a 0 0 0 0 0 0 0 0 0 0 0 0 0 0 0 0 0 0 0 0 0 0 0 0 0 0 0
c 0 0 0 0 0 0 0 0 0 0 0 0 0 0 0 0 0 0 0 0 0 0 0 0 0 0 0
h 0 0 0 0 0 0 0 0 0 0 0 0 0 0 0 0 0 0 0 0 0 0 0 0 0 0 0
i 0 0 0 0 0 0 0 0 0 0 0 0 0 0 0 0 0 0 0 0 0 0 0 0 0 0 0
n 0 0 0 0 0 0 0 0 0 0 0 0 0 0 0 0 0 0 0 0 0 0 0 0 0 0 0
e 0 0 0 0 0 0 0 0 0 0 0 0 0 0 0 0 0 0 0 0 0 0 0 0 0 0 0
s 0 0 0 0 0 0 0 0 0 0 0 0 0 0 0 0 0 0 0 0 0 0 0 0 0 0 0
  0 0 0 0 0 0 0 0 0 0 0 0 0 0 0 0 0 0 0 0 0 0 0 0 0 0 0
  0 0 0 0 0 0 0 0 0 0 0 0 0 0 0 0 0 0 0 0 0 0 0 0 0 0 0
  0 0 0 0 0 0 0 0 0 0 0 0 0 0 0 0 0 0 0 0 0 0 0 0 0 0 0
  0 0 0 0 0 0 0 0 0 0 0 0 0 0 0 0 0 0 0 0 0 0 0 0 0 0 0
  0 0 0 0 0 0 0 0 0 0 0 0 0 0 0 0 0 0 0 0 0 0 0 0 0 0 0
  0 0 0 0 0 0 0 0 0 0 0 0 0 0 0 0 0 0 0 0 0 0 0 0 0 0 0
  0 0 0 0 0 0 0 0 0 0 0 0 0 0 0 0 0 0 0 0 0 0 0 0 0 0 0
  0 0 0 0 0 0 0 0 0 0 0 0 0 0 0 0 0 0 0 0 0 0 0 0 0 0 0
  0 0 0 0 0 0 0 0 0 0 0 0 0 0 0 0 0 0 0 0 0 0 0 0 0 0 0
  0 0 0 0 0 0 0 0 0 0 0 0 0 0 0 0 0 0 0 0 0 0 0 0 0 0 0
  0 0 0 0 0 0 0 0 0 0 0 0 0 0 0 0 0 0 0 0 0 0 0 0 0 0 0
  0 0 0 0 0 0 0 0 0 0 0 0 0 0 0 0 0 0 0 0 0 0 0 0 0 0 0
MAX MACHINES IN A CELL
14
LENGTHS OF MACHINES
2.0 1.0 1.0 2.0 3.0 1.0 2.0 1.0 1.0 1.0 2.0 1.0 1.0 2.0 1.0 1.0 2.0 2.0 1.0 1.0 2.0 1.0 2.0 3.0 2.0 1.0 2.0
WIDTHS OF MACHINES
1.0 1.0 2.0 2.0 1.0 1.0 2.0 2.0 1.0 1.0 2.0 1.0 1.0 2.0 2.0 1.0 1.0 1.0 2.0 1.0 2.0 1.0 1.0 1.0 2.0 2.0 1.0
```

REFERENCES

Askin, R.G. and Chiu, K.S., (1990) A graph partitioning procedure for machine assignment and cell formation in group technology. *International Journal of Production Research*, **28**,(8), 1555–72.

Ballakur, A. and Steudel, H.J. (1987) Within-cell utilization based heuristic for designing cellular manufacturing system. *International Journal of Production Research*, **25** (5), 639–65.

Boe, W.J. and Cheng, C.H. (1991) A close neighbor algorithm for designing cellular manufacturing systems. *International Journal of Production Research*, **29**, (10) 2097–16.

Burbidge, J.L. (1963) Production flow analysis. *The Production Engineer*, **42** (12), 742.

Burbidge, J.L. (1977) A manual method of production flow analysis. *The Production Engineer*, **56**, (10) 34–38.

Carpenter, G.A. and Grossberg, S. [1788] ART of adaptive pattern recognition by self-organizing neural network. *Computer* **21**, (3) 77–88.

Chan, H.M. and Milner, D.A. (1982) Direct clustering algorithm for group formation in cellular manufacture. *Journal of Manufacturing Systems*, **1**(1), 65–75.

Chandrasekharan, M.P. and Rajagopalan, R. (1986) An ideal seed non-hierarchical clustering algorithm for cellular manufacturing. *International Journal of Production Research*, **24**(2), 451–64.

Chandrasekharan, M.P. and Rajagopalan, R. (1987) ZODIAC–an algorithm for concurrent formation of part-families and machine cells. *International Journal of Production Research*, **25**(6), 835–50.

De Beer, C. and De Witte, J. (1978) Production flow synthesis. *CIRP Annalen*, **27**, 389–92.

De Beer, C., Van Gerwen, R. and De Witte, J. (1976) Analysis of engineering production systems as a base for production-oriented reconstruction. *CIRP Annalen*, **25**, 439–41.

De Witte, J. (1980) The use of similarity coefficients in production flow analysis *International Journal of Production Research*, **18**(4) 503–14.

Heragu, S.S. (1992) Recent models and techniques for the layout problem, *European Journal of Operational Research*, **57**(2), 136–44.

Heragu, S.S. (1994) Group technology and cellular manufacturing. *IEEE Transactions on Systems, Man, and Cybernetics*, **SMC-24**(2) 203–15.

Heragu, S.S. and Alfa, A.S. (1990) A simulated annealing based approach to solve the facility layout problem, in *Proceedings of Fourth Advanced Technology Conference*, Washington DC, November 5–7, 489–99.

Heragu, S.S. and Alfa, A.S. (1992) Experimental analysis of simulated annealing based algorithm for the layout problem *European Journal of Operational Research*, **57**(2), 190–202.

Heragu, S.S. and Gupta, Y.P. (1994) A heuristic for designing cellular manufacturing facilities. *International Journal of Production Research*, **32**(1), 125–40.

Heragu, S.S. and Kakuturi, S.R. (1992) Grouping and placement of machine cells. DSES Technical Report 37–92–341, Rensselaer Polytechnic Institute, Troy, NY.

Iri, M. (1968) On the synthesis of loop and cutset matrices and related problems. *RAAG Memoirs*, **4**(A-XII), 376–410.

Kaparthi, S. and Suresh, N.C. (1992) Machine component cell formation in group technology – a neural network approach, *International Journal of Production Research*, **30**(6), 1353–68.

King, J.R. (1980) Machine-component grouping in production flow analysis: an approach using a rank order clustering algorithm. *International Journal of Production Research*, **18**(2), 213–32.

King, J.R. and Nakornchai, V. (1982) Machine-component group formation in group technology: review and extension. *International Journal of Production Research*, **20**(2), 117–33.

Kusiak, A. and Chow, W.S. (1987) An efficient cluster identification algorith, *IEEE Transactions on Systems, Man, and Cybernetics*, **SMC-17**(4), 696–99.

Logendran, R. (1990) A workload based model for minimizing total intercell and intracell moves in cellular manufacturing. *International Journal of Production Research*, **28**(5), 913–25.

McAuley, J., (1972) Machine grouping for efficient production, *The Production Engineer*, **51**(2), 53–57.

McCormick, W.T., Schweitzer, P.J. and White, T.E. (1972) Problem decomposition and data reorganization by a clustering technique. *Operations Research*, **20**, 993–1009.

Muther, R. (1973) *Systematic Layout Planning*, Cahners Books, Boston, MA.

Nagi, R., Harhalakis, G. and Proth, J. (1990) Multiple routings and capacity considerations in group technology applications. *International Journal of Production Research*, **28**(12), 2243–57.

Rajagopalan, R. and Batra, J.L. (1975) Design of cellular production systems – a graph theoretic approach. *International Journal of Production Research*, **13**(6), 567–79.

Rajamani, D., Singh, N. and Aneja, Y.P. (1992) A model for cell formation in manufacturing systems with sequence dependence. *International Journal of Production Research*, **30**(6), 1227–35.

Seifoddini, H. and Wolfe, P.M. (1986) Application of the similarity coefficient method in group technology. *IIE Transactions*, **22**(1), 271–77.

Srinivasan, G. and Narendran, T.T. (1991) GRAPHICS – a nonhierarchical clustering algorithm for group technology. *International Journal of Production Research*, **29**(3), 463–78.

Srinivasan, G., Narendran, T.T. and Mahadevan, B. (1990) An assignment model for the part-families problem in group technology. *International Journal of Production Research*, **28**(1), 145–52.

Teng, S. and Black, J.T., (1990) Cellular manufacturing systems modeling: the petri net approach. *Journal of Manufacturing Systems*, **9**(1), 45–54.

Vakharia, A.J. and Wemmerlov, U. (1990) Designing a cellular manufacturing system: a materials flow based operations sequence. *IIE Transactions*, **22**(1), 84–97.

Vohra, T., Chen, D., Chang, J.C. and Chen, H. (1990) A network approach to cell formation in cellular manufacturing. *International Journal of Production Research*, **28**(11), 2075–84.

Wemmerlov, U. and Hyer, N.L. (1986) Procedures for the part family/machine group identification problem in cellular manufacturing. *Journal of Operations Management*, **6**(2), 125–147.

CHAPTER TEN

Intelligent scheduling systems: an artificial-intelligence-based approach

Jigish S. Zaveri and Ali F. Emdad

10.1 INTRODUCTION

Current developments in the field of microelectronics and computing technology have made it possible to apply the results achieved in the field of artificial intelligence (AI) to developing systems that exhibit the attributes of AI. There are potentially numerous benefits that could be realized by incorporating the new trends evolving in AI to develop systems that could search for new heuristic sequencing/scheduling rules in complex manufacturing systems.

Scheduling problems are encountered in a wide variety of domains, including the manufacturing and service-based industries. In the broadest terms, scheduling involves the allocation of resources over a period of time to perform a collection of jobs, given the constraints of the system. In this sense, scheduling is a decision-making activity for determining a schedule. Discovering methodologies that will consistently generate 'satisfycing' if not optimal schedules, however, is a very difficult task. This is because most scheduling problems belong to a class of NP-complete combinatorial problems and hence are prime candidates for the application of AI technology.

This chapter discusses in detail a genetics-based machine learning system (GBMLS) that has been designed to search for useful sequencing/scheduling heuristics. The rest of the chapter is organized as follows. In section 10.2 we discuss the key issues involved in scheduling. Section 10.3 provides an overview of genetic algorithms (GAs) and highlights the advancements and diversity of research and practice in the area. In section 10.4 we present a detailed description of the experimental design. In section 10.5 we present the problem statement and the results of the experiment. In section 10.6 we discuss key features of the system and outline the directions for future research.

10.2 ISSUES INVOLVED IN SCHEDULING

A typical scheduling problem in the manufacturing industry can be stated as follows. There exists a manufacturing facility that consists of a set of machines, each of which can perform a variety of operations with varying efficiencies. Given a set of jobs, each of which requires a given set of operations to be performed in a prespecified order for its completion. The goal is to schedule all the operations required to process the given set of jobs on the given set of machines in such a way that the overall objective(s) of the system are met. Observe that the number of feasible schedules grows exponentially along each dimension, such as machines, tools and operations. As a simple example, the sequencing of ten jobs through three operations has $(10!)^3$ or more than 10^{19} possible schedules in a job shop without any alternative machines. The number of possible schedules grows still larger when alternative machines can perform different operations using different set-ups and with varying efficiencies.

The number of possible schedules or the solution space of possible schedules is bound by the constraints imposed by the system. Hence, identifying, representing and correctly and efficiently using the constraints is of utmost importance. Failure to identify a constraint can lead to infeasible schedules being generated, while addition of a false constraint can lead to inefficient schedules, or in extreme cases the system may not be able to generate a feasible schedule.

Constraints can be of many different types, and will be domain specific. In other words, each domain of the system environment will have its own unique set of constraints. Machine availability, machine capabilities (including types of operations, tool changes required to perform these operations, processing times, capacity limits), precedence requirements for the jobs, loading/unloading times and capacities, transportation facilities and capacities and personnel requirements are typical physical constraints. On the other hand, organizational goals like meeting due dates, balanced machine utilization, maximizing production, minimizing work-in-process and costs can also be viewed as constraints; while the goal of the system is to produce a feasible schedule that will satisfy the constraints. It should be noted that the physical constraints are *hard*, meaning they cannot be relaxed, while others are *soft*, meaning they can be relaxed at some cost. Additionally scheduling decisions must account for unexpected failures in some of the system components.

Given the above scenario, in most cases, the human scheduler generates schedules using simple heuristic algorithms that often results in poor system performance. In this chapter we demonstrate that our genetics-based scheduler is able to search sequencing/scheduling heuristics. This would result in considerable savings in terms of time and money. Further, after the system has discovered the scheduling rule(s), if (they) can be incorporated in the construction of a rule-based expert system. It should be noted that the sequence of releasing of jobs determines the schedule and hence we use the terms sequencing and scheduling interchangeably throughout the chapter.

10.3 AN OVERVIEW OF GENETIC ALGORITHMS

The technique of genetics-based machine learning draws on the seminal ideas of Holland (1975) about a class of algorithms called genetic algorithms (GAs). Numerous studies on the application of various kinds of GA-based approaches to computationally hard optimization problems from diverse domains include communications network configuration (Coombs and Davis, 1987; Davis and Coombs, 1987, 1989), gas and oil pipeline operations (Goldberg, 1983; Goldberg and Kuo, 1987), image registration (Grefenstette and Fitzpatrick, 1985), surveillance in warfare (Kuchinski, 1985), travelling salesman problem (Grefenstette *et al.* 1985; and Belew and Booker, 1991), multi-objective workforce scheduling (Hilliard *et al.*, 1989) and the time-constrained scheduling of limited resources (Syswerda and Palmucci, 1991).

In nature, a combination of natural selection and procreation permits the development of living species that are highly adapted to their environments. A GA is an algorithm that operates on a similar principle. When applied to a problem, this algorithm uses a genetics-based mechanism to iteratively generate new solutions from currently available solutions. It then replaces some or all of the existing members of the current solution pool with the newly created members. The motivation behind the approach is that the quality of the solution pool should improve with the passage of time, a process much like the 'survival of the fittest' principle that nature seems to follow.

More specifically, a GA operates on a population (of solutions) of fixed size (P) and iteratively performs three steps, in which it:

1. evaluates the **absolute fitness** of each individual in the population;
2. converts the absolute fitness values to **relative fitnesses** and selects individuals, based on their relative fitnesses, to populate a gene pool of size P;
3. uses various kinds of **genetic operators** on the gene pool to construct a new population of size P.

This three-step process is repeated until some prespecified stopping criteria are met. At this point, the population consists of a set of **highly fit** individuals. It is one or more of these highly fit individuals that we seek. For our purposes, each individual in the population corresponds to one possible solution to the scheduling problem of interest.

There are three basic kinds of genetic operators that are used by a GA. These are called **reproduction, crossover** and **mutation**. Usually, there is a predefined probability of procreation via each of these operators. Traditionally, these probability values are selected such that reproduction and crossover are the most frequently used, with mutation being resorted to only relatively rarely. This is because the mutation operator is a random operator and serves to introduce diversity in the population. The kind of operator to be applied to each member of the gene pool is determined by random choice based on these probabilities. Of the three operators, reproduction and mutation involve only a single parent and result in the creation of a single offspring. The standard crossover operator called a **simple crossover** has numerous variants such as partially-mapped, position-based, order-based, subtour chunking, cyclic, acyclic, inversion and edge-recombination crossovers. All of these involve two parents. Depending on operator and problem context, each generates either one or two offspring. For a detailed description and discussion of the various types of crossover operators, see Grefenstette *et al.* (1985); Oliver, Smith and Holland (1987); Goldberg (1989); Starkweather *et al.* (1991); Syswerda (1989); Syswerda and Palmucci (1991).

Genetic algorithms have been incorporated as part of two systems, LS1 (Learning System 1) (Smith, 1980) and Classifier Systems (Holland and Rietman, 1978; Booker, Goldberg and Holland 1989; Goldberg, 1989). Both systems are called 'rule discovery systems' and use the GA to facilitate the discovery of new rules to populate the knowledge base. The systems differ in their rule representation schemes and in their approach to fitness evaluation.

10.3.1 Advantages

There are many advantages to the GA approach. First, it is not a 'history independent' technique and hence is not just a random walk through the solution space. A GA uses the technique of **unsupervised induction** to discover new rules. Second, it allows simultaneous generation and evaluation of multiple sequences in parallel at each iteration. Additionally, in a GA, a parent sequence(s) is selected from a genetic pool with a probability that is in proportion to its relative fitness. Thus sequences with high performance have a higher probability of being selected as parents than their low performing counterparts. We now discuss how GAs utilize the technique of *unsupervised induction.*

A GA is an unsupervised learning technique because there is no external agent (i.e. supervisor or teacher) that periodically indicates to a system that embodies a GA, how much of what it has learned is correct or incorrect, useful or useless. We now discuss the ways in which inductive learning occurs in GAs.

First, unlike more conventional AI-based search methods such as A* search, and hill climbing, search using GAs does not depend on heuristic knowledge (for instance, a look-ahead function that estimates the cost-to-go from the current position) to prune a search space. That is, given a partial solution, traditional AI search methods choose to discontinue or pursue this solution path on the basis of estimates made about solution quality. These estimates are usually in the form of bounds on the quality of the complete solution if the current path were to be followed through to generate a complete solution. Hence, at each iteration or stage of the search process, conventional AI methods typically focus on the most promising path(s) from among several candidate paths, although the path itself may change as the search progresses.

On the other hand, search using GAs operates with complete solutions only. This method does not seek to abandon/pursue a partial solution based on anticipated quality. High quality solution(s) are used to generate other complete solutions of better quality. The emphasis is on simultaneous searches conducted from multiple locations within the solution space to generate many different (complete) solutions. Hence there is comparatively less emphasis on pruning out whole chunks of the solution space based on estimated quality.

The reason why quality improves over time is because, through sampling, a GA-based system learns by employing a technique called sample-based induction (Holland *et al.*, 1986). In selecting high quality solution(s) to generate subsequent solution(s), the GA approach incrementally modifies the contents of the knowledge system to improve performance. For example the GA approach progressively creates a population whose members all possess building blocks, schemata or patterns that render the subsequent solutions of high quality.

The technique of GAs views the solution space as being made up of *good* and *bad* regions. They make such inferences concerning the quality of entire regions in the solution space based on the available building blocks. Another name for sample-based induction is 'empirical induction' (Michalski and Kodratoff, 1990).

Induction in genetics-based systems may also be seen from a different perspective. For example the GA method permits the revision of population fitness values during the course of problem solving. The population fitness values determine the ordering or relative importance of the members of the current population. Consequently, this ordering impacts the composition of the gene pool and, thus, the selection of members for procreation activity.

For example, if the members are represented using 'If-Then' rules, a GA-based system would allow for changes in rule priorities over time. This is in sharp contrast with conventional expert systems where rule priorities are fixed for the duration of the consultation. In conventional expert systems, the associated rule priorities determine the order in which rules are considered for activation. For instance, in the simplest case, we could have a top-down priority scheme where the rules are considered in serial physical order from first to last. Thus a conventional expert system would activate or fire rules in exactly the same sequence based on some predefined priority scheme. This implies that if the system were asked to re-solve a problem, it would repeat exactly the same steps that it did in each of its previous attempts. Obviously, the firing of a rule at any stage of the problem-solving process does not result in a change of the priority value of any rule. In effect, the system has learned nothing new about modifying its problem processing ability with experience, with the intent to improve its performance.

In a GA-based system, on the other hand, the rule priorities are allowed to be dynamic and change during a problem-solving episode. Moreover, this change is not a random change and is the result of the sample-based induction process described earlier that seeks to improve the quality of available knowledge. This additional knowledge is available to the system during subsequent stages of the problem-solving exercise. Thus, if the system were asked to re-solve a problem, it may not repeat the same steps that it did in each of its previous attempts. Additionally, this could result in the system using the new knowledge acquired or generated over the course of one consultation session in subsequent consultations. Thus, in comparison to fixed-priority systems, we could expect a GA-based system to show superior monotonic improvement in performance at subsequent trials in re-solving the same problem.

10.4 SIMULATION DESIGN

The main purpose of these simulations is to study the learning mechanism and examine the efficacy of using GAs for job sequencing to develop good quality schedules for manufacturing systems. We first demonstrate that our GBMLS discovers the optimal SPT (shortest processing time) rule that minimizes the total flow time in a single machine case. Next we demonstrate that the GBMLS discovers the optimal Johnson rule that minimizes the total makespan in the two-machine case. The processing time for each of the jobs is generated from a uniformly distributed population of random numbers between 1 and 25 with the exception that no number could be repeated. This means that we ignore a number that has been generated earlier. This was to eliminate the possibility of alternative optimal solutions and thus ensure that there was only one optimal sequence for each of the two cases discussed.

The use of GAs requires that certain parameter settings be selected. The selection of these parameter settings influences the performance of the GA. Prior work by DeJong (1975) suggests parameter settings that have been used in a numerous applications of GAs. For our experiments we use the following parameter settings. The population size N is fixed at 50. The type of crossover operator used is subtour chunking (for a detailed description of this operator see Grefenstette *et al.*, 1985). This operator produces a child sequence by selecting chunks from the parent sequences. The probabilities of crossover and mutation are fixed at 0.6 and 0.001 respectively. The genetic pool was created by employing the elitist strategy. This strategy stipulates that the best sequence survives intact to the next generation. After insuring this, the genetic pool is created using the pure selection strategy where each sequence has copies in the pool in proportion to its observed performance. At each iteration one child sequence is inserted in the population and the relative fitness of the current members of the population is evaluated. Deletion of a member is then accomplished by removing the worst member of the population. Each GA generation involves the creation and evaluation of 50 child sequences. We allowed the GA to run through 2000 generations.

The general strategy being explored in this GBMLS application is as follows:

Initialization

Start with a set of N randomly generated 'seed' job sequences.

Iteration

1. Measure the quality of each job sequence currently available in terms of predefined measures. (Re)Assess the fitness value for each sequence currently available.
2. Create a genetic pool where each sequence has copies in proportion to its observed performance.
3. Depending on the operator probabilities, select a genetic operator. Depending on the type of genetic operator used randomly select one or two parent sequence(s) from the genetic pool.
4. Produce one child sequence and evaluate its performance in terms of the prespecified measure of performance.
5. Insert this new member in the population and re-evaluate the relative fitnesses of all members in the population. Delete the worst job sequence.
6. Repeat the above steps 3–5 until N child sequences have been created.
7. If total number of GA iterations = 2000, stop, else go to step 1.

The GBMLS just described was implemented using the C programming language on a Sequent Symmetry S81 machine equipped with 27 parallel processors and running DYNIX, a version of the UNIX operating system. Random numbers were generated using the DYNIX system function, 'random'. This function uses a nonlinear additive feedback random number generator employing a table of size 31 long integers to return successive pseudo random numbers in the range from 0 to $2^{(31-1)}$. The period of this random number generator is very large, approximately $16 \times 2^{(31-1)}$. 'Random' also allows one to produce a sequence of numbers that can be duplicated, if desired. This function is used at the following locations during the execution of the program:

1. generating the initial population;
2. selecting parents from the genetic pool;
3. selecting the type(s) of genetic operator to apply;
4. selecting the crossover location(s);
5. selecting the mutation location(s).

The function 'random' is initialized only once, at the beginning of the first replication for each of the two experiments described in sections 10.5.1 and 10.5.2. For each of the following 29 successive replications, the starting seed used was the one that was obtained at the end of the preceding replication. Due to the large periodicity of the random number generator, we would expect that the random starting population for each of the replications to be different. Even if the stream of $16 \times 2^{(31-1)}$ numbers was exhausted and the generator recycles, we believe that the invocation of the generator for a variety of tasks (as listed above), apart from starting population generation, would cause the seeds for different replications to be fairly disparate. We conducted random checks to verify the diversity of the starting populations by examining some of them visually for each of the experiments discussed.

10.5 EXPERIMENTS

10.5.1 Experiment 1

In the case of a single machine, and a given set of jobs, if we want to minimize total flow time (flow time is the total time each of the jobs spent in the system), then the optimal sequencing rule SPT states: 'Sequence the jobs in the increasing order of processing times.'

Problem statement
Given a single machine capable of performing one operation and ten jobs each requiring a single operation, with variable processing times, the objective of the resultant schedule is to minimize total flow time.

If we perform complete enumeration of all possible job sequences, the total number of sequences that will have to be evaluated (before we can select the optimum) is 10! or 3 628 800. Alternatively, we demonstrate that our GBMLS discovers the optimum sequence and hence the aforementioned sequencing rule by examining relatively far fewer sequences.

Table 10.1 gives the processing times for the ten jobs. It should be noted that the processing times include the set-up times. Using the optimal SPT sequencing we arrive at the following job sequence: J_6, J_2, J_{10}, J_7, J_8, J_5, J_4, J_1, J_3, J_9, with a total flow time of 507. The results for each of the 30 replications are given in Table 10.2. The first column shows the percentage of the total number of sequences generated to obtain the optimal sequence from the maximum possible number of sequences. The second column indicates the number of replications in which the optimal sequence was obtained. For example, in exactly three replications the GBMLS was able to obtain the optimal sequence by examining less than 0.2% of the maximum possible number of sequences. Also note that even in the worst case the GBMLS was required to examine only a maximum of 2.5% sequences before it could discover the optimal sequence.

Table 10.1 Job processing times (Experiment 1)

Job number	*Processing time*
J_1	18
J_2	5
J_3	21
J_4	17
J_5	14
J_6	2
J_7	8
J_8	12
J_9	24
J_{10}	7

Table 10.2 Results of Experiment 1

Number of replications	*Percentage of sequences examined*
3	0.2
9	0.5
20	1.0
25	1.5
28	2.0

10.5.2 Experiment 2

Given a set of jobs that have to be processed on two machines, Johnson's algorithm arrives at an optimal schedule that minimizes makespan which is measured as the time elapsed from commencement of processing on the first job to the completion of processing of the final job in the schedule. Johnson's rule is stated as follows.

- Let x_i be the processing time of job i on the first machine and let y_i be the processing time of job i on the second machine.
- Examine all the x_i and y_i values and select the minimum value: if that value is an x_i, then the ith job should be scheduled in the first available position; if the y_i value is the minimum, then that job should be scheduled in the last available position.
- Remove that job from the list of jobs to be scheduled.
- Examine the remaining x_i and y_i values and select the minimum: if the second value is also an x_i, then that job should be scheduled in the second place; if the y_i value is the minimum, then that job should be scheduled as the last job (if the first job also had an lower y_i value, then the second job should be sequenced in the second last position).
- Proceed to sequence the remaining unscheduled jobs in the same manner – by searching for the minimum x_i, y_i value and placing the job at the first available position at the beginning or towards the end of the schedule depending on whether it has a lower x_i or lower y_i value.

To make the algorithm clear, here is an example that has been worked out.

Problem statement
Given two machines each capable of performing one operation and ten jobs each requiring two operations on the two machines, with different processing times. The objective of the resultant schedule is to minimize makespan.

Table 10.3 gives the processing times for ten jobs on two machines. Applying Johnson's rule the resultant optimal job sequence and/or schedule is derived as follows. Examining the processing times, we observe that the minimum value is 1 for job J_6 on machine 1; hence job J_6 is placed at the beginning of the sequence. The next lower value is 3 for job J_1 and it is on the second machine; hence job J_1 is placed at the end of the sequence. The next lower value is 4 for job J_4 on machine 1; hence job J_4 is placed at the second position in the sequence. Among the jobs not placed in the sequence, the next lower value is 7 for job J_2 on machine 2 and hence job J_2 is placed in the second last position. Proceeding thus, we get the optimal sequence as $J_6, J_4, J_9, J_5, J_{10}, J_7, J_3, J_8, J_2, J_1$. The makespan associated with this optimal sequence is 141.

Again, in the absence of Johnson's rules, we would have to evaluate 362 880 (10!) number of schedules, before we can determine an optimum schedule. Alternatively, we demonstrate that our GBMLS discovers the optimum sequence and hence the aforementioned sequencing rule by

Table 10.3 Job processing times (Experiment 2)

Job number	*Processing time* *Machine 1*	*Machine 2*
J_1	20	3
J_2	13	7
J_3	25	10
J_4	4	21
J_5	11	16
J_6	1	5
J_7	22	19
J_8	15	8
J_9	9	14
J_{10}	18	23

examining relatively far fewer sequences. The results of this experiment are given in Table 10.4, showing as before the percentage of the total number of sequences generated to obtain the optimal sequence from the maximum possible number of sequences, and the number of replications in which the optimal sequence was obtained. For example, in exactly three replications the GBMLS was able to obtain the optimal sequence by examining less than 0.2% of the maximum possible number of sequences. Also note that even in the worst case the GBMLS was required to examine only a maximum of 2.0% sequences before it could discover the optimal sequence.

Table 10.4 Results of Experiment 2

Number of replications	*Percentage of sequences examined*
3	0.2
7	0.5
18	1.0
27	1.5
30	2.0

10.6 CONCLUSIONS

In both of the above experiments we demonstrated that the GBMLS successfully discovered the optimal sequence and hence the underlying optimal rule by examining less than 2.5% of the total number of possible job sequences. However, both the cases examined were simple examples of a manufacturing environment. We would also like to examine the performance of our system for other complex manufacturing environments where optimal sequencing rules do not exist. For example, we would like to examine the performance of our system for the case of flexible manufacturing systems where each of the machines can perform more than one operation, and a given operation can be performed by more than one machine.

A second concern is that the parameter settings for the GA affects the performance of the system. Future research would be geared towards identifying optimal parameter settings for the GA, such as population size, type and frequency of usage of genetic operators, in order to optimize the performance of the system.

ACKNOWLEDGEMENTS

The authors are grateful for the support from the David D. Lattanze Center for Executive Studies on Information Systems.

REFERENCES

Belew, R.K. and Booker, L.B. (eds) (1991) *Proceedings of the Fourth International Conference on Genetic Algorithms*, University of California, San Diego. Morgan Kaufmann Publishers, San Mateo, CA.

Booker, L.B., Goldberg, D.E. and Holland, J.H. (1989) Classifier systems and genetic algorithms. *Artificial Intelligence*, no. 40, 235–282.

Coombs, S. and Davis, L. (1987) Genetic algorithms and communication link speed design: constraints and operators, in *Proceedings of the Second International Conference on Genetic Algorithms*, Lawrence Erlbaum Associates, Hillsdale, NJ, pp. 257–260.

Davis, L. and Coombs, S. (1987) *Genetic algorithms and communication link speed design: theoretical considerations*, in *Proceedings of the Second International Conference on Genetic Algorithms*, Lawrence Erlbaum Associates, Hillsdale, NJ, pp. 252–256.

Davis, L. and Coombs, S. (1989) Optimizing network link sizes with genetic algorithms, in *Modelling and Simulation Methodology: Knowledge Systems Paradigms* (eds M. Elzas, T. Oren and B.P. Zeigler) North Holland, Amsterdam.

DeJong, K.A. (1975) *Analysis of the behavior of a class of genetic adaptive systems.* PhD Dissertation, Department of Computer and Communication Sciences, University of Michigan.

Goldberg, D.E. (1983) *Computer-aided gas pipeline operation using genetic algorithms and rule learning*. PhD Dissertation, College of Engineering, University of Alabama.

Goldberg, D.E. (1989) Genetic algorithms in search, optimization, and machine learning, Addison-Wesley, Reading, MA.

Goldberg, D.E. and Kuo, C.H. (1987) Genetic algorithms in pipeline optimization. *Journal of Computing in Civil Engineering*, **1**, 128–141.

Grefenstette, J.J. and Fitzpatrick, J.M. (1985) Genetic search with approximate function evaluations, in *Proceedings of an International Conference on Genetic Algorithms and Their Applications*, Lawrence Erlbaum Associates, Hillsdale, NJ, pp. 112–120.

Grefenstette, J.J., Gopal, R., Rosmaita, B. and Van Gucht, D. (1985) Genetic algorithms for the travelling salesman problem, in *Proceedings of the First International Conference on Genetic Algorithms and Their Applications*, Lawrence Erlbaum Associates, Hillsdale, NJ, pp. 160–168.

Hilliard, M.R., Liepins, G.E., Rangarajan, G. and Palmer, M. (1989) Learning decision rules for scheduling problems: a classifier hybrid approach, in *Proceedings of the Sixth International Conference on Machine Learning*, Morgan Kaufmann Publishers, San Mateo, CA, pp. 188–200.

Holland, J.H. (1975) *Adaptation in Natural and Artificial Systems*, The University of Michigan Press, Michigan.

Holland, J.H. and Reitman, J.S. (1978) Cognitive systems based on adaptive algorithms, in *Pattern Directed Inference Systems* (eds D.A. Waterman, and F. Hayes-Roth), Academic Press, New York.

Holland, J.H., Holyoak, K.J., Nisbett, R.E. and Thagard, P.R. (1986) *Induction: Processes of Inference, Learning, and Discovery*, The MIT Press, Cambridge, MA.

Kuchinski, M.J. (1985) *Battle management systems control rule optimization using artificial intelligence. Technical Report No. NSWC MP 84–329*, Naval Surface Weapons Center, Dahlgren, VA.

Michalski, R.S. and Kodratoff, Y. (1990) Research in machine learning: recent progress, classification of methods, and future directions, in *Machine Learning: An Artificial Intelligence Approach* (eds Y. Kodratoff, and R.S Michalski), Vol. 3, Morgan Kaufmann Publishers, San Mateo, CA.

Oliver, I., Smith, D. and Holland, J. (1987) A study of permutation crossover operators on the travelling salesman problem, in *Proceedings of the Second International Conference on Genetic Algorithms and Their Applications*, Lawrence Erlbaum Associates, Hillsdale, NJ, pp. 224–230.

Smith, S.F. (1980) *A learning system based on genetic adaptive algorithms*. PhD Dissertation, Department of Computer Science, University of Pittsburgh.

Starkweather, T., McDaniel, S., Mathias, K., Whitley, D. and Whitley, C. (1991) A comparison of genetic sequencing operators, in *Proceedings of the Fourth International Conference on Genetic Algorithms* (eds R.K. Belew and R.L. Booker), University of California, San Diego. Morgan Kaufmann Publishers, San Mateo, CA, pp. 69–76.

Syswerda, G. (1989) *Uniform crossover in genetic algorithms*, in *Proceedings of the Third International Conference on Genetic Algorithms* (ed. D. Schaffer), Morgan Kaufmann Publishers, San Mateo, CA, pp. 67–72.

Syswerda, G. and Palmucci, J. (1991) The application of genetic algorithms to resource scheduling, in *Proceedings of the Fourth International Conference on Genetic Algorithms* (eds R.K. Belew and R.L. Booker), University of California, San Diego. Morgan Kaufmann Publishers, San Mateo, CA. pp. 502–508.

CHAPTER ELEVEN

Optimizing assembly time for printed circuit board assembly

Haomin Li and Ratnesh Kumar

11.1 INTRODUCTION

A major challenge for any manufacturing industry is to increase its productivity, safety and quality without significantly increasing the per unit cost of production. In recent years, **automated** and **flexible** manufacturing (Desrochers, 1990; Jurgen, 1987) systems have received considerable attention from design engineers and other researchers, as they provide an alternative to a higher productivity, increased safety and quality of production. Robots have become an integral part of many of the manufacturing systems, replacing humans in workplaces, thereby not only increasing the safety of production but also the productivity and quality of production.

In many manufacturing settings, robots are employed to perform a sequence of assembly operations, each of which requires a fixed amount of time. In order to maintain a high rate of productivity, a robot must complete a given sequence of assembly operations in a minimal possible time. For example, in an electronic assembly machine (EAM), which is used for automated assembly of components in surface mount technology or insertion pin technology, a pick-and-place robot is employed to pick up components from a set of pickup slot locations and to place them on the desired locations of the printed circuit board (PCB) being assembled. Some of the reasons why automated assembly machines are becoming increasingly popular in industrial settings are as follows.

- Dense boards and tiny surface mount components make manual assembly very difficult, if not impossible.
- Greater product mix and smaller volume of production require automated assembly, as there is not enough time for acquiring skills for manual assembly.

- Automated assembly provides consistent quality.
- Automated assembly results in increased rate of production.

A typical EAM that is manufactured by Quad Systems Incorporated, Horsham, PA, and its workplace arrangement is shown in Fig. 11.1 (Quad, 1987). This machine is installed in the Surface Mount Technology Laboratory of the Center for Robotics and Manufacturing at the University of Kentucky.

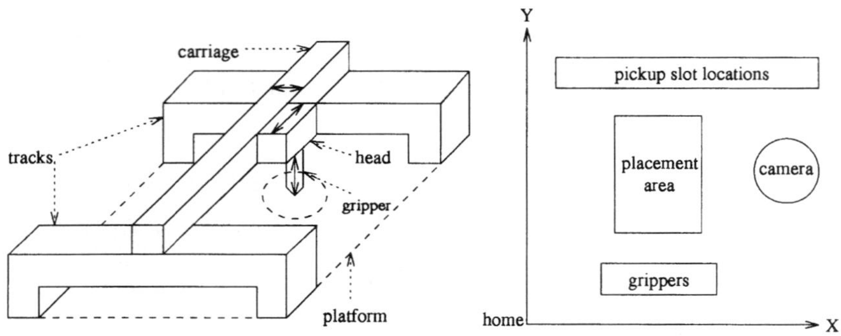

Fig. 11.1 An EAM and its workplace arrangement.

The machine has a **carriage** with a **pick-and-place head** on it. The carriage moves horizontally on **tracks** in, say, the **X**-direction, and the head moves along the carriage in a perpendicular horizontal (say, *Y*-) direction. These *X*- and *Y*-direction movements can occur concurrently. The **gripper** (also called the **chuck**) on the head can move in the vertical *Z*-direction for performing pickups and placements; it can also rotate around the vertical axis for performing the proper alignment of components. Thus such a robot essentially has three translational joints and one rotational joint. Components to be assembled are kept in various component feeders around the PCB. Components of relatively smaller sizes and regular shapes are supplied by **reel** and/or **vibration** feeders. In these cases, components are picked from a single fixed pickup slot location of the feeder, while the slot location is continually supplied with new components as they are picked and assembled. Components of relatively large sizes and irregular shapes are supplied simply on component **trays**. Thus each such component has its own pickup slot location.

The sequence of operations performed by such a pick-and-place robot can be described as follows (Quad, 1987). The robot starts from its designated **home location**, moves to a pickup slot location, gets a component, then moves to the desired placement location on the PCB being assembled and places it there. After placement, the robot moves to possibly another pickup slot location, gets another component and repeats the prior sequence of operations. In case components are not of the same size, the robot also changes its gripper by moving to the gripper location during the assembly operation. Also, certain critical components are tested for proper alignment against an upward-looking camera, in which case after picking such critical components and prior to their placements, the robot moves to the camera location. After completing the assembly operation, the robot returns to its home location and waits for the next PCB to be assembled to arrive. Figure 11.1 depicts a typical workplace arrangement of such an EAM.

It is clear that the overall assembly time of a given PCB depends on two decision variables:

1. the assignment of the pickup slot locations (in general several alternative pickup slot locations are available);
2. the sequence in which the components are assembled.

The problem of determining an optimal sequence of assembly in PCB assembly has been addressed by many researchers. In Chan and Mercier (1989), it is formulated as an instance of the traveling salesman problem; Ball and Magazine (1988), study it as an instance of the directed postman problem; in Downs, Linn and Martinson (1992), heuristics such as the 'typewriter' method and 'S-shape' method – which are commonly used in practice for determining the sequence of assembly – are compared to a proposed heuristic called the 'block algorithm' method. The problem of optimal slot assignment has recently been addressed in Sadiq, Landers and Taylor (1993), in which several heuristics are proposed for assignment and reassignment of slots so as to minimize the set-up time. However, as noted above, the two problems are interdependent. We have developed a unified approach to address both these optimization issues.

Several other issues related to PCB assembly have been addressed in the literature: in Johari (1992), the design of an entire PCB assembly line of a given capacity and flowtime has been studied; Bao (1988) proposes an expert system to assess the relative importance of various design factors in PCB manufacturing; in Maimon and Shtub (1991) methods of determining group of PCBs and set of components to be assembled on each PCB of such a group by a single EAM (when several concurrently operating EAMs are available), so as to optimize the throughput of production, have been obtained; Lacksonen and Joshi (1991) address the problem of minimization of the number of components to be manually assembled arising from interference with previously assembled components; in Spuinski, Egbelu and Lehtihet (1991) an algorithm for sequencing of adhesive, solder and component disposition in PCB assembly has been presented; Chang and Terwilliger (1987) present an expert-system-based, computer-aided, process planner for planning a PCB assembly process; in

Randhawa, McDowell and Faruqui (1985) an integer-programming-based technique has been developed for determining an optimal sequencer mix in PCB assembly; BoothRoyd and Shinohara (1986) present a survey of component insertion times at various industrial sites; in Cunningham and Browne (1986) a heuristic search-based strategy is proposed for scheduling different PCB jobs on different assembly machines to maximize overall throughput; and Taylor and Graves (1990) present a simulation-based comparison of different routing strategies for optimizing the production rate in PCB assembly.

The work of several control theorists (e.g. Akella, Choong and Gershwin, 1984; Akella and Kumar, 1986; Chase and Ramage, 1992; Dallery, David and Xie, 1989; Gershwin, 1989; Kouikoglou and Phillis, 1991; Kumar and Seidman, 1990; Lu and Kumar, 1991; Perkins and Kumar, 1989; Sharifnia, 1988; Towsley, Sparaggis and Cassandras, 1992) that deals with issues of stability, throughput/average delay optimization, etc. in a general manufacturing system are also of interest, as they are also applicable to PCB manufacturing systems.

The rest of the chapter is organized as follows. In section 11.2, we develop the relevant notation as well as describe the problem. In section 11.3, we present two different formulations of the problem, both based on an integer-programming approach (Minoux, 1986; Nemhauser and Wolsey, 1988). Since integer programming is an NP-complete problem (Garey and Johnson, 1979), we obtain polynomially computable, near-optimal feasible solutions using the first formulation of the problem. This requires techniques for solving the minimum weight-matching problem (Papadimitriou and Steglitz, 1982) and those for solving the traveling salesman problem (Lauler *et al.*, 1985). These near-optimal solutions provide upper bounds for the overall assembly time. The second formulation establishes that the optimization problem is an instance of linear integer programming (Nemhauser and Wolsey, 1988). So a linear programming relaxation of it can be used to obtain a polynomially computable lower bound for the overall assembly time. In section 11.4, we provide the simulation results of our work, and finally, section 11.5 concludes the work presented.

11.2 NOTATIONS AND PROBLEM STATEMENT

In this section we introduce the necessary notations and describe the optimization problem for an EAM of the type shown in Fig. 11.1. As described above, due to differences in sizes of components, in general different grippers are needed to complete the assembly of a given PCB. However, since a significant amount of time is wasted during interchange of grippers, a gripper, once selected, is not released until all the components that it can assemble have been assembled. The optimization problem is that of determining an assignment of pickup slots and a sequence of assembly of components for each of the grippers independently, since:

1. the time spent during the interchange of grippers is independent of the sequence in which grippers are selected;
2. each gripper assembles components of different sizes so that the set of possible pickup slot locations for each of the grippers is mutually exclusive.

Hence, it may be assumed, without loss of generality, that exactly one gripper is used for assembly. In the subsequent analysis we make this assumption.

Let $\mathbb{N}$ denote the set of all natural numbers, and $\mathbb{R}_+$ denote the set of positive real numbers. Let $n \in \mathbb{N}$ be the number of components to be assembled on a given PCB. Since many of the components of the same **make** are picked from the same pickup slot location, the number $\bar{n} \in \mathbb{N}$ of different pickup slot locations occupied during assembly is smaller than n. Let the **type function** $\tau : \{1, \ldots, n\} \to \{1, , \bar{n}\}$ identify components with identical pickup slot locations, i.e., for $i, j \leq n$, $\tau(i) = \tau(j)$ if and only if components i and j are picked from the same pickup slot location. It must be emphasized that the type function does not always identify components of the same make to be of the same type, as some of the components, although of the same make, are picked from different pickup locations.

Given a component type $l \in \{1, \ldots, \bar{n}\}$, $\tau^{-1}(l) := \{i \leq n | \tau(i) = l\}$ identifies all components of type l. Let $m \in \mathbb{N}$ be the total number of available pickup slot locations. Since each component of the same type is assigned the same pickup slot location, we must have $\bar{n} \leq m$. Let $[t_{ij}]_{n \times m} \in \mathbb{R}_+^{n \times m}$ be a matrix such that for each $i \leq n$ and $j \leq m, t_{ij} \in \mathbb{R}_+$ denotes the **travel time** from the component location i to the pickup slot location j. The travel times may or may not include any set-up times such as component pickup times, component placement times, and in case of critical components, camera visit times.

Thus an instance of assembly time optimization problem when a single Quad type machine is available consists of numbers n, $\bar{n}$, $m \in \mathbb{N}$, the type function $\tau : \{1, \ldots, n\} \to \{1, \ldots, \bar{n}\}$, and the travel time matrix $[t_{ij}]_{n \times m} \in \mathbb{R}_+^{m \times n}$ as described above. In order to simplify future notation, given a set S, we use $|S|$ to denote its cardinality. Also, give an optimization problem P, *val*(P) is used to denote the optimal value of P.

11.3 OPTIMIZATION USING INTEGER PROGRAMMING

In this section we provide two alternative integer-programming formulations of the optimization problem introduced in the previous section. Since an instance of an integer-programming problem is an NP-complete problem, polynomially computable, near-optimal feasible solutions are obtained by further constraining the first formulation so that it reduces to two different subproblems for which known solution techniques exist. These near-optimal feasible solutions provide an upper bound for the overall assembly time. Polynomially computable lower bounds for the overall assembly time can be obtained by applying any of the known relaxation techniques to the second formulation which poses the optimization problem as an instance of **linear** integer-programming problem (Minoux, 1986).

11.3.1 First formulation of optimization problem

In this subsection we present a *quadratic* integer-programming-based formulation of the optimization problem, which is further constrained to obtain two different well known subproblems:

1. an instance of the minimum weight-matching problem for determining an assignment of pickup slot locations;
2. an instance of the traveling salesman problem for determining the sequence of assembly.

Let $[u_{ik}]_{n\times n} \varepsilon \{0,1\}^{n\times n}$ be an $n\times n$-dimensional binary-valued decision variable with the interpretation that for each $i,\ k \le n,\ u_{ik} = 1$ if and only if component k is assembled immediately after the assembly of component i. Also, $[v_{jl}]_{m\times \bar{n}} \varepsilon \{0,1\}^{m\times \bar{n}}$ be an $m \times \bar{n}$–dimensional binary-valued decision variable with the interpretation that for each $j \le m,\ l \le \bar{n},\ v_{jl} = 1$, if and only if pickup slot location j contains the components of type l. Clearly, for each $i \le n,\ j \le m,\ k \le n$ there is a motion from the ith component location to the kth component location visiting the jth pickup slot location if and only if $u_{ik} = v_{j\mathcal{T}(k)} = 1$, in which case a travel time of $[t_{ij} + t_{kj}]$ is spent. Thus the optimatization problem can be formulated as:

P1
$$\min_{\substack{[u_{ik}]\varepsilon\{0,1\}^{n\times n} \\ [v_{jl}]\varepsilon(0,1)^{m\times\bar{n}}}} \left[\sum_{i\le n} \sum_{j\le m} \sum_{k\le n} (t_{ij} + t_{kj}) u_{ik} v_{j\mathcal{T}(k)} \right]$$

subject to

A1.
$$\sum_{k\le n} u_{ik} = 1, \forall i \le n;$$

A2.
$$\sum_{i\le n} u_{ik} = 1, \forall k \le n;$$

A3. $$\sum_{i\varepsilon S}\sum_{k\varepsilon S} u_{ik} \leq |S| - 1, \forall S \subset \{1, \ldots, n\};$$

A4. $$\sum_{j\leq m} v_{jl} = 1, \forall l \leq \bar{n};$$

A5. $$\sum_{l\leq \bar{n}} v_{jl} \leq 1, \forall j \leq m.$$

Note that u_{ik} represents the number of motions – either zero or one – from component location i to component location k, i.e., $u_{ik} = 1$ if and only if component location k is visited immediately after visiting component location i. Since a sequence in which component locations must be visited constitutes a **Hamiltonian cycle** (Lawler *et al.*, 1985), the decision variables $[u_{ik}]_{n\times n}$ must satisfy the corresponding constraints A1–A3 (Lawler *et al.*, 1985). Constraint A1 requires that there must be exactly one motion exiting each component location; constraint A2 requires that there must be exactly one motion entering each component location; constraint A3 requires that for each proper subset $S \subset \{1, \ldots, n\}$ of component locations, the sequence of motions restricted to S must not form a cycle. This constraint is also known as the subtour elimination constraint. Constraint A4 requires that each type of component must be assigned to exactly one pickup slot location, which implies that all components of the same type are picked from the same pickup slot location. Constraint A5 requires that each pickup slot location must contain at most one type of component.

We next provide a method for obtaining near-optimal feasible solutions for P1 using the known solution techniques of the minimum weight-matching problem and the traveling salesman problem. The basic idea is to further constrain P1 so that it reduces to two subproblems, each of which depends only on a single set of decision variables – either $[u_{ik}]_{n\times n}$ or $[v_{jl}]_{m\times\bar{n}}$. Thus the complexity of the problem is significantly reduced. Consider the objective function of P1:

$$\sum_{i\leq n}\sum_{j\leq m}\sum_{k\leq n}(t_{ij} + t_{kj})u_{ik}v_{j\tau(k)} = \left(\sum_{i\leq n}\sum_{j\leq m}\sum_{k\leq n} t_{ij}u_{ik}v_{j\tau(k)}\right) + \left(\sum_{i\leq n}\sum_{j\leq m}\sum_{k\leq n} t_{kj}u_{ik}v_{j\tau(k)}\right). \tag{11.1}$$

The first sum term represents the total 'up-loading time', i.e., it is the time spent in moving from a component location to a pickup slot location, whereas the second sum term represents the total 'down-loading time', i.e., it is the time spent in moving from a pickup slot location to a component location. The second sum term in the objective function can be simplified as follows:

$$\sum_{i\leq n}\sum_{j\leq m}\sum_{k\leq n} t_{ij}u_{ik}v_{j\tau(k)} = \sum_{j\leq m}\sum_{k\leq n}\sum_{i\leq n} t_{kj}u_{ik}v_{j\tau(k)} = \left[\sum_{j\leq m}\sum_{k\leq n} t_{kj}v_{j\tau(k)}\left(\sum_{i\leq n} u_{ik}\right)\right] \tag{11.2}$$

$$= \sum_{j\leq m}\sum_{k\leq n} t_{kj}v_{j\tau(k)} \tag{11.3}$$

$$= \sum_{j\leq m}\sum_{k\varepsilon\tau^{-1}(l)}\sum_{l\leq\bar{n}} t_{kj}v_{j\tau(k)}$$

$$= \sum_{j\leq m}\sum_{l\leq\bar{n}}\sum_{k\varepsilon\tau^{-1}(l)} t_{kj}v_{j\tau(k)}$$

$$= \sum_{j\leq m}\sum_{l\leq\bar{n}}\sum_{k\varepsilon\tau^{-1}(l)} t_{kj}v_{\tau jl} \tag{11.4}$$

$$= \left[\sum_{j\leq m}\sum_{l\leq\bar{n}} v_{jl}\left(\sum_{k\varepsilon\tau^{-1}(l)} t_{kj}\right)\right]$$

$$= \sum_{j\leq m}\sum_{l\leq\bar{n}} \bar{t}_{jl}v_{jl}, \tag{11.5}$$

where equality of (11.2) follows from the fact that $t_{kj}v_{j\tau(k)}$ is a constant with respect to the summation variable i; equality of (11.3) follows from A2, equality of (11.4) follows from the fact that if $k\varepsilon\tau^{-1}(l)$, then $\tau(k) = l$, and

$$\bar{t}_{jl} := \sum_{k\varepsilon\tau^{-1}(l)} t_{kj};\ \forall j \leq m, \forall l \leq \bar{n}. \tag{11.6}$$

Thus $\bar{t}_{jl}$ is the total time spent in moving from pickup slot location j to all component locations of type l.

Let $[v^{*}_{jl}]_{m\times n}\varepsilon\{0,1\}^{m\times\bar{n}}$ denote an assignment that minimizes the total 'down-loading time' $\sum_{j\leq m}\sum_{l\leq\bar{n}} t_{jl}v_{jl}$, i.e.,

$$[v^{*}_{jl}]_{m\times\bar{n}} := arg\left[\min_{\substack{[v_{jl}]\varepsilon\{0,1\}^{m\times\bar{n}}\\ subjecttoA4-A5}}\left(\sum_{j\leq m}\sum_{l\leq\bar{n}} \bar{t}_{jl}v_{jl}\right)\right]. \tag{11.7}$$

Then for the optimization problem P1 we have:

$$\min_{\substack{[u_{ik}]\epsilon\{0,1\}^{n\times n},[v_{jl}]\epsilon\{0,1\}^{m\times\bar{n}} \\ \text{subject to } A1-A5}} \left[\sum_{i\leq n} \sum_{j\leq m} \sum_{k\leq n} (t_{ij} + t_{kj}) u_{ik} v_{j\mathcal{T}(k)} \right]$$

$$= \min_{\substack{[u_{ik}\epsilon\{0,1\}^{n\times n},[v_{jl}]\epsilon\{0,1\}^{m\times\bar{n}} \\ \text{subject to } A1-A5}} \left[\left(\sum_{i\leq n} \sum_{j\leq m} \sum_{k\leq n} t_{ij} u_{ik} v_{j\mathcal{T}(k)} \right) + \left(\sum_{j\leq m} \sum_{l\leq \bar{n}} \bar{t}_{jl} v_{jl} \right) \right] \tag{11.8}$$

$$\leq \min_{\substack{[u_{ik}]\epsilon\{0,1\}^{n\times n} \\ \text{subject to } A1-A5}} \left[\left(\sum_{i\leq n} \sum_{j\leq m} \sum_{k\leq n} t_{ij} u_{ik} v^*_{j\mathcal{T}(k)} \right) + \left(\sum_{j\leq m} \sum_{l\leq \bar{n}} \bar{t}_{jl} v^*_{jl} \right) \right] \tag{11.9}$$

$$= \min_{\substack{[u_{ik}]\epsilon\{0,1\}^{n\times n} \\ \text{subject to } A1-A5}} \left[\left(\sum_{i\leq n} \sum_{j\leq m} \sum_{k\leq n} t_{ij} u_{ik} v^*_{j\mathcal{T}(k)} \right) + \left(\sum_{j\leq m} \sum_{l\leq \bar{n}} \bar{t}_{jl} v^*_{jl} \right) \right] \tag{11.10}$$

$$= \min_{\substack{[u_{ik}]\epsilon\{0,1\}^{n\times n} \\ \text{subject to } A1-A5}} \left[\sum_{i\leq n} \sum_{k\leq n} u_{ik} \left(\sum_{j\leq m} t_{ij} u_{ik} v^*_{j\mathcal{T}(k)} \right) \right] + \sum_{j\leq m} \sum_{l\leq \bar{n}} \bar{t}_{jl} v^*_{jl} \tag{11.11}$$

$$= \min_{\substack{[u_{ik}]\epsilon\{0,1\}^{n\times n} \\ \text{subject to } A1-A5}} \left[\sum_{i\leq n} \sum_{k\leq n} t^*_{ij} u_{ik} \right] + \sum_{j\leq m} \sum_{l\leq \bar{n}} \bar{t}_{jl} v^*_{jl},$$

where the equality of (11.8) follows from (11.1) and (11.5); the inequality of (11.9) follows from the fact that minimization is performed on a smaller domain; the equality of (11.10) follows from the fact that constraints A4–A5 apply to decision variables $[v_{jl}]_{m\times n}$ which are no longer present in the objective function; the equality of (11.11) follows from the fact that the second sum term is a constant; and

$$t^*_{ik} := \sum_{j\leq m} t_{ij} v_{j\mathcal{T}(k)}, \forall i, k \leq n, \tag{11.12}$$

where $[v^*{}_{jl}]_{m\times n}$ is defined in (11.7). Thus $t^*{}_{ik}$ is the total time spent in moving from component location i to the pickup slot location that contains components of the same type as component k, under the pickup slot assignment of (11.7).

It follows from the above analysis that the optimization problem P1 can be constrained to obtain two different subproblems – an instance of the traveling salesman problem (TSP) and an instance of the minimum weight-matching problem (MWMP):

TSP. $$\min_{\substack{[u_{ik}]\epsilon\{0,1\}^{n\times n} \\ \text{subject to } A1-A5}} \left[\sum_{i\leq n} \sum_{k\leq n} t^*_{ij} u_{ik} \right]; t^*_{ik} \text{ defined in (11.12)};$$

MWMP.
$$\min_{\substack{[v_{jl}]\varepsilon\{0,1\}^{m\times\bar{n}} \\ \text{subject to } A1-A5}} \left[\sum_{j\leq m}\sum_{l\leq\bar{n}} \bar{t}^*_{jl} v_{jl}\right] ; \bar{t}^*_{jl} \text{ defined in (11.6) ;}$$

Note that constraints A1–A3 correspond to the traveling salesman problem constraints (Lawler *et al.*, 1985), and constraints A4–A5 correspond to the minimum weight-matching problem constraints (Papadimitriou and Steglitz, 1982)

We have proved the following result.

Theorem 11.1

Consider the optimization problems P1, TSP and MWMP defined above. Then

$$val(P1) \leq [val(\text{TSP}) + val(\text{MWMP})].$$

Thus $[val(\text{TSP}){+}val(\text{MWMP})]$ provides an upper bound for the optimal assembly time. However, it does not directly provide a polynomially computable upper bound, as although an instance of the minimum weight-matching problem is of polynomial complexity (Lawler, 1976; Papadimitriou and Steglitz, 1982) an instance of the traveling salesman problem is NP-complete (Garey and Johnson, 1977). Fortunately, there exist several polynomial time algorithms, such as nearest neighbor, nearest insertion, 2-optimality and 3-optimality (Lawler *et al.*, 1985), which yield near-optimal feasible solutions and provide an upper bound for the TSP. Thus using these algorithms it is possible to obtain a polynomially computable upper bound for the optimization problem considered here.

11.3.2 Second formulation of optimization problem

In this subsection we establish that the optimization problem is an instance of *linear* integer programming (Nemhauser and Wolsey, 1988). Thus techniques and software tools such as GAMS (Brooke, Kendrick and Meerhaus, 1988) for solving linear integer-programming problems can be directly applied. In particular, polynomially computable lower bounds for the overall assembly time can be easily obtained by using any of the relaxation techniques such as linear-programming relaxation and lagrangian relaxation (Nemhauser and Wolsey, 1988).

Let $[\pi_{ijk}]_{n\times m\times n}\varepsilon\{0,1\}^{n\times m\times n}$ be an $n \times m \times n$–dimensional binary-valued decision variable with the interpretation that for each $i \leq n$, $j \leq m$, $k \leq n$, $\pi_{ijk} = 1$ if and only if after placing the ith component, the pick-and-place robot picks a component from the jth pickup slot location and places it at the kth component location. Clearly, a travel time of $[t_{ij} + t_{kj}]$ is spent if and only if $\pi_{ijk} = 1$. Thus the optimization problem can be formulated as:

P2. $$\min_{[\pi_{ijk}]\varepsilon\{0,1\}^{n\times m\times n}} \left[\sum_{i\le n} \sum_{j\le m} \sum_{k\le n} (t_{ij} + t_{kj})\pi_{ijk} \right]$$

subject to

$$x_{ik} := \sum_{j\le m} \pi_{ijk}, y_{jk} := \sum_{i\le n} \pi_{ijk};$$

B1. $$\sum_{k\le n} x_{ik} = 1, \forall i \le n;$$

B2. $$\sum_{i\le n} x_{ik} = 1, \forall k \le n;$$

B3. $$\sum_{i\varepsilon S} \sum_{k\varepsilon S} x_{ik} \le |S| - 1, \forall S \subset \{1, \dots, n\};$$

B4. $$y_{jk'} = y_{jk}; \forall k' \varepsilon \tau^{-1}(\tau(k)), \forall k \le n, \forall j \le m;$$

B5. $$\left[\sum_{l\le \bar{n}} \left(\frac{1}{\tau^{-1}(l)} \sum_{k\varepsilon\tau^{-1}(l)} y_{jk} \right) \right] \le 1; \forall j \le m.$$

Since π_{ijk} represents the number of motions – either zero or one – from ith component location to kth component location that also visits the jth pickup slot location, x_{ik} represents the total number of motions from ith component location to kth component location that visit *any* pickup slot location. Hence $x_{ik} = 1$ if and only if component location k is visited immediately after visiting component location i. As noted in the previous subsection, a sequence in which component locations must be visited constitutes a Hamiltonian cycle. Hence x_{ik} must satisfy the corresponding constraints B1–B3 (Lawler *et al.*, 1985). (These constraints are exactly the same as constraints A1–A3 of the formulation given in the previous subsection.)

Constraints B4–B5 are needed to ensure that components of the same type are picked from the same slot location, and each slot location contains at most one type of component. y_{jk} represents the total number of motions from jth pickup slot location to kth component location. Note that for each $k \le n$

$$\sum_{j\le m} y_{jk} = \sum_{j\le m} \sum_{i\le n} \pi_{ijk} = \sum_{i\le n} \sum_{j\le m} \pi_{ijk} = \sum_{i\le n} x_{ik} = 1, \tag{11.13}$$

which implies that $y_{jk} \le 1$ for all $j \le m$ and $k \le n$. Constraint B4 requires that a component location k is visited from a pickup slot location j if and only if all component locations with components of the same type as component k are visited from jth pickup slot location (which implies that components of the same type are picked from the same pickup slot location). Constraint B5 requires that for each $j \leq m$, the number of motions from pickup slot location j to component locations with components of *different* types must not exceed one (which implies that each pickup slot location must contain at most one type of components).

It is clear that P2 is an instance of a linear integer-programming problem (Minoux, 1986; Nemhauser and Wolsey, 1988) for which techniques of polynomial complexity for computing near-optimal feasible solutions are well known. Such solutions provide upper bounds for the overall assembly time. Since it is computationally intractable to obtain an optimal solution of the optimization problem, it is impractical to obtain a *direct* measure of closeness of near-optimal solutions to an optimal one. Instead it is desirable to obtain polynomially computable lower bounds for the overall assembly time: the gap between lower and upper bounds provides an *indirect* measure of closeness of near-optimal solutions to an optimal one.

An additional advantage of the linear integer-programming formulation of the optimization problem is that polynomially computable lower bounds for the overall assembly time can be easily obtained from it. Several algorithms of polynomial complexity are available in the literature for obtaining a lower bound for a given linear integer-programming problem (Nemhauser and Wolsey, 1988). In particular, a linear-programming relaxation of P2, called LP, is obtained by letting the decision variables $[\pi_{ijk}]_{n\times m\times n}$ take values in the $n \times m \times n$–dimensional unit interval $[0,1]_{n\times m\times n} \supset \{0,1\}_{n\times m\times n}$. Clearly,

$$val(\text{LP}) \leq val(\text{P2}).$$

The technique of Lagrangian relaxation (Nemhauser and Wolsey, 1988) can also be considered to obtain a lower bound.

Next we prove that P2 is equivalent to P1.

Theorem 11.2

The optimization problems P1 and P2 as defined above are equivalent.

Proof: We first show that an instance P1 can be reduced to an instance of P2. For each $i \leq n, j \leq m, k \leq n$, define $\pi_{ijk} := u_{ik}v_{j\tau(k)}$. With this definition, the objective function of P1 reduces to that of P2. Also,

$$x_{ik} := \sum_{j\leq m} \pi_{ijk} = \sum_{j\leq m} u_{ik}v_{j\tau(k)} = u_{ik}\sum_{j\leq m} v_{j\tau(k)} = u_{ik},$$

where the last equality follows from constraint A4. Since for each $i,k \leq n, u_{ik}$ satisfies constraints A1–A3, it follows that x_{jk} satisfies constraints B1–B3 as desired.

Next consider

$$y_{jk} := \sum_{i \leq n} \pi_{ijk} = \sum_{i \leq n} u_{ik} v_{j\tau(k)} = v_{j\tau(k)} \sum_{i \leq n} u_{ik} = v_{j\tau(k)},$$

where the last equality follows from constraint A2. Since $y_{jk} = v_{j\tau(k)}$, it follows that if $k' \leq n$ is such that $k' \varepsilon \tau^{-1}(\tau(k))$, i.e., $\tau(k') = \tau(k)$, then

$$y_{jk'} = v_{j\tau(k')} = v_{j\tau(k)} = y_{jk},$$

which shows that constraint B4 also holds. It remains to show that constraint B5 also holds. Since constraint A5 holds, in order to show that B5 holds it suffices to show for each $j \leq m$ and $l \leq \bar{n}$

$$\frac{1}{|\tau^{-1}(l)|} \sum_{k \varepsilon \tau^{-1}(l)} y_{jk} = v_{jl}, \tag{11.14}$$

Since $k \varepsilon \tau^{-1}(l)$, we have $\tau(k) = l$. Hence $y_{jk} = v_{j\tau(k)} = v_{jl}$. Thus

$$\begin{aligned} \frac{1}{|\tau^{-1}(l)|} \sum_{k \varepsilon \tau^{-1}(l)} y_{jk} &= \frac{1}{|\tau^{-1}(l)|} \sum_{k \varepsilon \tau^{-1} l} v_{jl} = \frac{1}{|\tau^{-1}(l)|} v_{jl} \left(\sum_{k \varepsilon \tau^{-1}(l)} 1 \right) \\ &= \frac{1}{|\tau^{-1}(l)|} v_{jl}(|\tau^{-1}(l)|) = v_{jl}, \end{aligned} \tag{11.15}$$

as desired.

Next we show that given an instance of P2, it can be reduced to an instance of P1. For each $i, k \leq n, l \leq$, define

$$u_{ik} := x_{ik} = \sum_{j \leq m} \pi_{ijk} \text{ and } v_{jl} := y_{jkl} = \sum_{i \leq n} \pi_{ijkl},$$

where $k_l \leq n$ is such that $\tau(k_l) = l$. Note that it follows from B4 that if $k', k'' \leq n$ are such that $\tau(k') = \tau(k'')$, then $y_{jk'} = y_{jk''}$; this implies that v_{jl} is unambiguously defined. We need to show that with the above definition of u_{ik} and v_{jl} the objective function of P2 reduces to that of P1, i.e., we need to prove that

$$\pi_{ijk} = u_{ik} v_{j\tau(k)} = x_{ik} y_{jk},$$

where the last equality follows from definitions of u_{ik} and v_{jl}. We have

$$x_{ik}y_{jk} = \left(\sum_{j'\leq m} \pi_{ij'k}\right)\left(\sum_{i'\leq n} \pi_{i'jk}\right) = \sum_{j'\leq m}\sum_{i'\leq n} \pi_{ij'k}\pi_{i'jk} = \pi_{ijk},$$

where the last equality follows from constraint B2, which states that for each $k \leq n$, there exists exactly one particular $i \leq n$ and one particular $j \leq m$ such that $\pi_{ijk} = 1$. It remains to show that constraints A1–A5 also hold.

Since $u_{ik} = x_{ik}$ and constraints B1–B3 hold, it is immediate that constraints A1–A3 hold. In order to see that constraint A4 holds, consider

$$\sum_{j\leq m} v_{jl} = \sum_{j\leq m} y_{jk_l} = \sum_{j\leq m}\sum_{i\leq n} \pi_{ijk_l} = \sum_{i\leq n}\sum_{j\leq m} \pi_{ijk_l} = \sum_{i\leq n} x_{ik_l} = 1,$$

where $k_l \leq n$ is such that $\tau(k_l) = l$, and the last equality follows from B2. This proves that constraint A4 holds. Finally, since constraint B5 holds, in order to show that constraint A5 holds, it suffices to show that for each $j \leq m$ and $l \leq \bar{n}$ equality of (11.4) holds. It follows from the definition of v_{jl} that $y_{jk} = v_{j\tau(k)}$, and since $k \varepsilon \tau^{-1}(l), \tau(k) = l$. Hence $y_{jk} = v_{jl}$ if $k \varepsilon \tau^{-1}(l)$. Thus equality of (11.4) can be derived using a derivation similar to that of (11.5).

The following corollary is an immediate consequence of Theorems 11.1 and 11.2.

Corollary 11.3

Consider optimization problems P1, TSP, MWMP, P2, LP as defined above. Then

$$val(\text{LP}) \leq val(\text{P2}) = val(\text{P1}) \leq [val(\text{TSP}) + val(\text{MWMP})].$$

11.4 SIMULATION RESULTS

Above we have proposed two different methods for optimizing assembly time of a given PCB: the first one is based on the traveling salesman problem (for determining sequence of assembly) and minimum weight matching (for determining assignment of pickup slots); the second is based on linear integer programming (for determining sequence of assembly as well as assignment of pickup slots). In this section we describe the way in which we have implemented the two proposed methods, present some simulation results and compare the effectiveness of our approach with other approaches that are currently in use in industries.

In industrial settings, two different approaches, namely, the typewriter method and S-shape method, are commonly used for determining sequence of assembly; and a 'greedy assignment' algorithm is commonly used for determining the assignment of pickup slots (Garey and Johnson, 1979). Figure 1.2 depicts the two methods for determining sequence of assembly. Both the typewriter and S-shape methods use relative x–y-coordinate location of each of the components to determine the sequence of assembly, and no information about the intercomponent distances is used. In both these methods, components with a smaller y-coordinate are assembled before components with a higher y-coordinate. In the typewriter method, components of the same y-coordinate are always assembled in an increasing order of their x-coordinates; however, in the S-shape method they are assembled in an increasing order of their x-coordinates and a decreasing order of their x-coordinates, alternately.

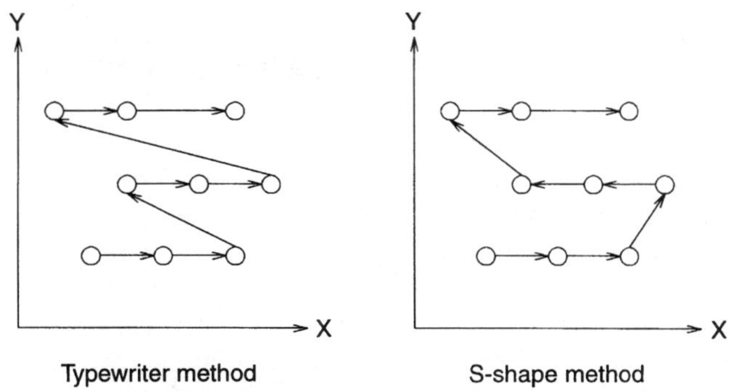

Fig. 11.2 Two commonly used sequencing methods.

The greedy assignment algorithm (GA) for determining the assignment of pickup slot locations works simply as follows. Firstly, the component types are prioritized according to their numbers. Thus given two types $l_1 \leq \bar{n}$ and $l_2 \leq \bar{n}$, l_1 is given priority in assignment of slots over l_2 if and only if the number of components of type l_1 is more than the number of components of type l_2, i.e., if $|\tau^{-1}(l_1)| > |\tau^{-1}(l_2)$. Next, the types with higher priority are assigned the nearest available slot location. In other words, if type $l \leq \bar{n}$ has the highest priority, then it is assigned to that pickup slot location $j \leq m$ for which the value $\bar{t}_{jl}$ as defined in (11.6) is minimum. (Recall that $\bar{t}_{jl}$ is the total down-loading time of all the components of type l when assigned the pickup slot location j.)

Next we describe the technique that we use for the implementation of our proposed approach. We use a commercially available software package called SAS/OR for obtaining a solution of the minimum weight-matching problem (MWMP) of the previous section, which determines the assignment of pickup slots. (This software is available on the University of Kentucky Computing Center (UKCC) machine.) Since an instance of MWMP is polynomially solvable, we are able to optimally solve this subproblem. We have developed our own software package called UKTSP for obtaining a near-optimal solution of the traveling salesman problem (TSP) of the previous section, which determines the sequence of assembly. We use well known heuristics such as nearest neighbor, nearest insertion, furthest insertion and random generation to generate an initial assembly sequence; and we use other sets of heuristics such as 2-optimality, 3-optimality and or-optimality to improve upon the initially generated assembly sequence. (Refer to Lawler *et al.* (1985) for a detailed description of these heuristics.) This software has been developed on our UNIX-based Engineering Computing Center (ECC) machine, and is encoded in the C programming language.

The lower bound for the optimal assembly time is obtained by implementing the second formulation of the problem. We apply the *General Algebraic Modeling System* (GAMS) software (Brooke, Kendrick and Meerhaus, 1988), which is available on UKCC, to solve the linear integer-programming formulation P2. However, in this case we are unable to exploit the inherent structure of the problem. In particular, we need to deal with order* $O(2^n)$ of subtour elimination constraints of B3. This exponential growth of the number of constraints makes it infeasible to consider all the subtour elimination constraints, and we relax this constraint by eliminating the most commonly occurring subtours – those of size two. Thus the lower bound we obtain is less tight than that obtainable by the LP relaxation of P2. However by adding 'higher order' subtour elimination constraints it is possible to make the lower bound tighter.

A random number generation program available on an ECC machine is used for generating uniformly distributed component location coordinates as well as pickup slot location coordinates. We require that the *x–y*-coordinates of component locations satisfy the constraint that $10 \leq x, y \leq 26$, which represents a PCB of size 16 × 16. Similarly, we require that *x–y*-coordinates of the pickup slot locations satisfy the constraint that $0 \leq x, y \leq 32$ and $10 \geq x, y \geq 26$, which represents a 32 × 32 workplace area, the central 16 × 16 area of which corresponds to the placement area, and the remaining represents the pickup area. We assume the following for simplicity.

1. No two components are of the same type, i.e., for each $l \leq \bar{n}, |\tau^{-1}(l)| = 1$, which implies that $\bar{n} = n$.

* Function $f(.)$ is said to be of order $g(.)$, denoted $O(g(.))$, if there exists a constant $c \varepsilon \mathbb{R}_+$ such that $f(.) \leq cg(.)$

2. The number of possible pickup slot locations is the same as the number of types of components, i.e., $m = \bar{n} = n$.
3. The travel time t_{ij} from component location $i \leq n$ and pickup slot location $j \leq m = n$ equals the Euclidean distance between them.

These assumptions are made only for ease in simulation, and do not reflect any restriction on the scope of our algorithms or software tools.

Table 11.1 summarizes simulation results for ten different test cases for $n = 100$. The first column lists the test number; the second column lists the seed value for random number generation; the third column lists the assembly time using the typewriter method (TW) together with greedy assignment algorithm (GA); the fourth column lists the assembly time using S-shape (SS) method together with the greedy assignment algorithm; the fifth column lists the assembly time using our approach, TSP together with MWMP; the sixth column lists the percentage improvement over the typewriter method together with the greedy assignment algorithm; and finally the seventh column lists the percentage improvement over the S-shape method together with greedy assignment algorithm.

Table 11.1 Savings resulting from TSP + MWMP formulation

No.	*Seed*	*TW + GA*	*SS + GA*	*TSP + MWMP*	*Reduction (%) with respect to*	
					TW+GA	*SS+GA*
1	1045	3394	3479	2570	24.28	26.13
2	42	3136	3177	2413	23.05	24.05
3	523	3260	3200	2375	27.15	25.78
4	205	3339	3368	2563	22.83	23.90
5	100	3280	3366	2665	18.75	20.83
6	2414	3324	3391	2469	25.72	27.19
7	51	3061	3139	2183	28.68	30.46
8	7125	3453	3524	2621	24.09	25.62
9	8	3193	3237	2455	23.11	24.16
10	99	3238	3320	2479	23.44	25.33
				Average:	24.11	25.35

These simulation results suggest that our proposed approach can considerably improve the assembly time (up to 25%).

Table 11.2 provides a measure of closeness of the near-optimal solutions developed using TSP and MWMP to the optimal ones by considering the **gap** between the upper and lower bounds, which is defined as:

$$\text{gap} := \left[\frac{\text{upper bound} \; - \; \text{lower bound}}{\text{upper bound}}\right] \times 100\%.$$

Table 11.2 Closeness of the proposed optimization algorithm to an optimal one

n	*Seed*	*Upper bound*	*Lower bound*	*Gap(%)*
10	12	281	279	0.71
20	497	502	499	0.59
30	3686	732	730	0.27
40	361	1050	1029	2.00
50	812	1313	1219	7.16
60	9252	1689	1532	9.29

Since the number of decision variables in P2 is of order $O(n^3)$ (as opposed to of order $O(n^2)$ in P1), we obtain the gap for six different cases where n ranges between 10 and 60 (as opposed to $n = 100$). The first column lists the value of n selected; the second column lists the seed value for random number generation; the third column lists the upper bound obtained using the proposed approach of TSP and MWMP; the fourth column lists the lower bound value obtained using the approach described above; the last column lists the gap.

The simulation results suggest that the proposed optimization approach produces solutions that are less that 2% away from the optimal ones when the value of n is small (within 40). However as the value of n increases the chances of the presence of subtours of size large than two increases, and a very conservative lower bound is obtained resulting in higher gap values. One should include higher order subtour elimination constraints to get a faithful lower bound, which, however, will require more powerful computing resources.

11.5 CONCLUSION

In this chapter we have presented two different formulations of the assembly time optimization problem in printed circuit board assembly. The overall assembly time for a given PCB depends upon the assignment of pickup slots as well as sequence of assembly. We have demonstrated that the overall optimization problem is an instance of the linear integer programming problem, and that the techniques for solving well known combinatorial optimization problems such as the minimum weight-matching problem and traveling salesman problem can be successfully applied to determine near-optimal solutions for the optimization problem. Simulation results show a consistent assembly time saving of 25% over some of the techniques currently in use in industrial settings. Also, the proposed approach is of polynomial time complexity $O(n^3)$, and provides solutions that are close to the optimal ones.

However, while running the simulation, some simplified assumptions were made, so that the improvements reflected by the simulations may not conform to a real world setting. We are currently in the process of implementing our algorithms on the placement machines located in the ECAT division of LEXMARK International Incorporation. This will give us a better idea of resulting improvements in assembly time.

ACKNOWLEDGEMENTS

We would like to thank our colleagues Prof. C. Lee of the Department of Mathematics, Prof. L.E. Holloway of the Department of Electrical Engineering and Mr Rod Heard of the Center for Robotics and Manufacturing for useful discussions during the course of this work.

REFERENCES

Akella, R. and Kumar, P.R. (1986) Optimal control of production rate in a failure prone manufacturing system. *IEEE Transactions on Automatic Control*, **31**(2), 116–26.

Akella, R., Choong, Y. and Gershwin, S.B. (1984) Performance of hierarchical production scheduling. *IEEE Transactions on Components, hybrids, and Manufacturing Technology*, **7**(3), 225–39.

Ball, M.O. and Magazine, M.J. (1988) Sequencing of insertion in printed circuit board assembly. *Operations Research*, **36**, 192–201.

Bao, H.P. (1988) An expert system for SMT printed circuit board design for assembly. *Manufacturing Review*, **1**(4), 275–80.

BoothRoyd G. and Shinohara, T. (1986) Components insertion times for electronic assembly. *International Journal of Advanced Manufacturing Technology*, **1**(5), 3–18.

Brooke, A. Kendrick, D. and Meerhaus, A. (1988) *GAMS, A User's Guide*, Scientific Press, Redwood City, CA.

Chan, D. and Mercier, D. (1989) IC insertion: an application of the traveling salesman problem. *International Journal of Production Research*, **27**(10), 1837–41.

Chang T. and Terwilliger, J. (1987) A rule based system for printed wiring assembly process planning. *International Journal of Production Research*, **25**(10) 1465–82.

Chase, C. and Ramadge, P.J. (1992) On real time scheduling policies for flexible manufacturing systems. *IEEE Transactions on Automatic Control*, **37**(4) 491–96.

Cunningham, P. and Browne, J. (1986) A LISP-based heuristic scheduler for automatic insertion in electronic assembly. *International Journal of Production Research*, **24**(6), 1395–1408.

Dallery, Y. David, R. and Xie, X. (1989) Approximate analysis of transfer lines with unreliable machines and finite buffers. *IEEE Transactions on Automatic Control*, **34** (9), 943–53.

Desrochers, A.A., (ed) (1990) *Modeling and control of Automated Manufacturing Systems*, IEEE Computer Society Press, Washington, DC.

Downs, K.A., Linn, R.J. and Martinson, N.L. (1992) Sequencing methods for single in-line packages in pcb assembly. *Manufacturing Review*, **5**(3), 175–83.

Garey, M.R. and Johnson, D.S. (1979) *Computers and Intractability: A Guide to the Theory of NP-Completeness*, W.H. Freeman and Company.

Gershwin, S.B. (1989) Hierarchical flow control: a framework for scheduling and planning discrete events in manufacturing systems, *Proceedings of IEEE*, **77**(1) 195–209.

Johari, P.K. (1992) Engineering a circuit board assembly line for a desired capacity and flowtime. *Journal of Manufacturing Systems*, **10**(6) 492–500.

Jurgen, R.K. (ed.) (1987) *Computers and Manufacturing Productivity*, IEEE Press, New York.

Kouikoglou, V.S. and Phillis, Y.A. (1971) An exact discrete-event model and control policies for production lines with buffers. *IEEE Transactions on Automatic Control*, **36**(5) 515–27.

Kumar, P.R. and Seidman, T.I. (1970) Dynamic instabilities and stabilization methods in distributed real-time scheduling of manufacturing systems. *IEEE Transactions on Automatic Control*, **35**(3), 289–98.

Lacksonen T. and Joshi, S. (1971) A minimum manual component insertion algorithm for PCB assembly based on graph theory. *Journal of Manufacturing Systems*, **9**(4), 345–52.

Lawler, E.L. (1976) *Combinatorial Optimization: Networks and Matroids*, Saunders College Publishing.

Lawler, E.L., Lenstra, J.K., Rinnooy Kan, A.H.G. and Shmoys, D.B. (eds.) (1985) *The Traveling Salesman Problems*, John Wiley and Sons.

Lu, S.H. and Kumar, P.R. (1971) Distributed scheduling based on due dates and buffer priorities *IEEE Transactions on Automatic Control*; **36**(12), 1406–16.

Maimon, O. and Shtub, A. (1971) Grouping methods for printed circuit board assembly. *International Journal of Productions Research*, **29**(7), 1379–1390.

Minoux, M. (1986) *Mathematical Programming: Theory and Algorithm*, John Wiley and Sons.

Nemhauser, G.L. and Wolsey, L.A. (1988) *Integer and Combinatorial Optimization*, John Wiley and Sons,

Papadimitriou, C.H. and Steiglitz, K. (1982) *Combinatorial Optimization: Algorithms and Complexity*, Prentice-Hall.

Perkins, J.R. and Kumar, P.R. (1989) Stable, distributed, real time scheduling of flexible manufacturing/assembly/disassembly systems. *IEEE Transactions on Automatic Control* **34**(2), 139–48.

Quad (1987) *QuadStar 100 Manual*, October, Quad Systems Corporation, Horsham, PA.

Randhawa, S.U., McDowell, E.W. and Faruqui, S. (1985) An integer programming application to solve sequencer mix problems in printed circuit board production, *International Journal of Production Research*, **23**(3), 543–53.

Sadiq, M. Landers, T.L. and Taylor, G.D. (1993) A heuristic algorithm for minimizing total production time for a sequence of jobs on a surface mount placement machine. *International Journal of Production Research*, **31**(6), 1327–41.

Sharifnia, A. (1988) Production control of manufacturing systems with multiple machine states. *IEEE Transactions on Automatic Control*, **33**(7), 620–26.

Spuinski, M.R. Egbelu, P.J. and Lehtihet, E.A. (1991) Automatic plan and robot code generation for PCB assembly. *Manufacturing Review*, **4**(3), 214–24.

Taylor, D. and Graves, R.J. (1990) An examination of routing flexibility for small batch assembly of printed circuit boards, *International Journal of Production Research*, **28**(11), 2117–35.

Towsley, D. Sparaggis, P.D. and Cassandras, C.G. (1992) Optimal routing and buffer allocation for a class of finited capacity queuing systems. *IEEE Transactions on Automatic Control*, **37**(9), 1446–51.

CHAPTER TWELVE

On integration of statistical process control and engineering process control: a neural network application

Fahimeh Rezayat

12.1 INTRODUCTION

In recent years the importance of quality has become increasingly apparent. Quality control in manufacturing has moved from detecting non-conforming products through inspection to continuously reducing variability in product performance and production process. Two existing fields which have been contributing to quality are **statistical process control** (SPC) and **engineering process control** (EPC). In the past, these fields have been developed in relative isolation from one another, resulting in process control engineers trained primarily in classical control theory and almost no training in statistics, data analysis and process noise control. On the other hand, statisticians have a poor understanding of process dynamics and classical control theory, but they have excellent training in the analysis of discrete data, the design of experiments and the methods for empirical process modeling (Gupta and Kumar, 1991). Both methodologies aim to bring all the process levels to their targets with minimum variability. Statistical process control commonly assumes that under normal conditions the process variability is driven by **common causes**, the effect of which is impossible or too expensive to reduce, and that an increase in variability or unfavorable changes in the process mean are due to **special causes**. The main aim of control is rapid detection and elimination of such special causes or, where possible, neutralization of their effect by an appropriate adjustment. The primary tools used to achieve this goal are control charts (Yashchin, 1993).

For many processes, engineers are able not only to describe mathematically their in-control behavior, but also the way the processes go out of control. This, in turn, enables one to take systematic corrective actions, based on the observed data related to raw materials, process and product parameters etc., that will keep the resulting process in a state of equilibrium with small variance. These actions are typically described by a mathematical formula and are frequently implemented by using automatic regulators and similar devices. Within this framework, the quality of control is measured in terms of the output variance or a similar loss function (Yashchin, 1993).

There has been a growing realization that both of these approaches will make a contribution, that both should be implemented in a complementary fashion and that quality will be gained through appropriate process adjustment (using EPC techniques) and through elimination of root of variability (using SPC techniques) see for example Vander Wiel *et al.*, 1992; Faltin *et al.*, 1993; Box 1993; Yashchin 1993. This chapter continues in that direction with one critical difference: we assume that no knowledge about the exact process dynamics exists.

When knowledge about the physical mechanism of a process is available, it sometimes allows one to deduce the correct form of the functional relationship linking the mean value of the output to the levels of control factors (input values) and construct a mathematical model which provides feedback to the system. When the necessary physical knowledge of the system is absent or incomplete, the relationship is traditionally approximated over the limited ranges of the experimental variables of input by a function, e.g. a polynomial (Box and Draper, 1987).

We believe that representing the underlying structure of a process by a mathematical relation that does not closely approximate the process, especially when the process is nonstationary and/or nonlinear, will limit the task of quality improvement. This fact has motivated the use of **neural networks** (NNs) in the field of engineering process control and in this study.

Since the decision-making process in advanced manufacturing systems has been growing difficult for humans, decision support methodologies have become widely adopted to assist in the process. Artificial intelligence (AI) is a class of decision support methodologies which is based on sophisticated computer programs and is used in manufacturing systems (Udo, 1992). There is a wide use of several types of AI in all aspects of manufacturing processes. A neural network is a type of AI effective for complex processes and processes that have underlying structure changes or are ill-structured. A neural network is able to recognize patterns and relationships among prevailing factors which are not apparent to human experts (Udo, 1992).

In the last few years, many studies have reported successful applications of NN methodology in many areas including strategic planning, resource allocation, scheduling, maintenance and repair, process control and planning of manufacturing. Some applications of

NN for process control are given in Tulunay (1991), Narendra and Parthasarathy (1990), Nguyen and Widrow (1990), Smith and Dagli (1991), Udo (1992), Chryssolouris, Lee and Domroese (1991). There is also a report on using NNs for quality control and machine vision (Udo, 1992). All the above methods use an NN as a controller for process variations reduction, and require either an explicit mathematical model or a second NN as an emulator for the underlying process.

This study simulates the performance of NN-based adaptive control proposed by Spall and Cristion (1992, 1994) to reduce the variation in viscosity of a polymer (Vander Wiel *et al.* 1992) around its target value under the assumption that no knowledge about the underlying dynamics of the process is available. The NN-based adaptive control is novel in that it employs only one NN as a controller and does not need information about the underlying nature of the process or a second NN to model the process. The study also employs the **exponentially weighted moving average (EWMA) chart**, a commonly used chart for **stochastic process** monitoring (Yashchin, 1993), and the **special cause control (SCC) chart** (Alwan and Roberts, 1988) to monitor the polymer's viscosity deviations from the target value.

Spall and Cristion estimate the NN weights of the approach by using the output error along with the **simultaneous perturbation stochastic approximation (SPSA)** algorithm (Spall, 1992). Since their method does not assume any specific representation of the underlying process, it can readily adapt to the changes in the process, and it seems a promising approach for improving the output quality of unknown and/or nonlinear dynamic processes.

The goals of this study are

1 to integrate statistical process control and engineering process control techniques in order to improve the quality of simulated polymers;
2 to examine the speed of EWMA and SCC charts in signaling the existence of a special cause;
3 to test how fast an SPSA-based NN controller readjusts the level of viscosity of the polymer within an acceptable variation range.

The remaining part of the chapter is organized as follows. Section 12.2 reviews briefly the structure of NN, the feedforward NN controller based on SPSA and statistical control charts. Section 12.3 provides simulation experiments and analysis of their findings. Section 12.4 presents conclusions and remarks.

12.2 PRELIMINARIES

12.2.1 A review of neural networks

A neural network is a system arranged in patterns similar to biological neural nets and modeled after the human brain. The main processing elements of an NN are individual neurons, analogous to the brain's neurons. These artificial neurons receive and sum 'information' from other neurons or external input stimuli, perform a transformation on the inputs and then pass on the transformed 'information' to other neurons or external outputs, the human brain is thought to work in a similar way. The 'information' passed from neuron to neuron can be thought of as a way to activate a response from certain neurons based on the information or stimulus received.

Neural networks possess certain strengths not found in conventional models (Udo, 1992). These strengths lead many to believe that the methodology will cause drastic changes in the way we process information. The most important characteristic of NN is that they can learn to simulate any behavior and can be used to generate the action necessary to produce a given response, i.e., to control a plant. The potential advantage of NNs is that, compared to mathematical simulation models, they can be trained using observed data only with no knowledge of the internal structure of the system or of modeling techniques.

Like other simulation models, NN substitute for the real system in predicting or controlling system responses for the purpose of dynamic control. Unlike many other methods, NN learning can be continuous and can automatically adapt to change in characteristics of a system.

In general, to design an NN, one has to determine the structure (number of nodes and layers) and iteratively estimate its parameters (weights). The process of estimating parameters of an NN is called training. The goal of NN weight estimation is to find a vector of weights that results in a sequence of control values where the control values, in turn, drive the process output close to the target value. Figure 12.1 presents the structure of a simple NN where its input layer has two nodes and output layer has one node and the two hidden layers have four and two nodes respectively. At an iteration the input to a layer node is the sum of the weighted output of the nodes of the previous layer and a bias weight. The output of a node is a nonlinear (linear) function of that node inputs.

Let $I_j^1(k)$ and $o_j^1(k)$ respectively denote the input and output of the jth node in the first layer, and $0_l^0(k)$ the output of the lth node of input layer at the kth iteration for the NN in Fig. 12.1. Then the input and output of the nodes for the first layer are calculated as:

$$I_j^1(k) = \sum_1^2 \theta_{jl}^1(k) o_l^0(k) + \text{ bias weight}, \quad j = 1, 2, \ldots, 4 \qquad (12.1)$$

$$o_j^1(k) = s(I_j^1(k)), \quad j = 1, 2, \ldots, 4 \qquad (12.2)$$

where $s(.)$ is a sigmoidal function and θ_{jl}^1 is the connection weight

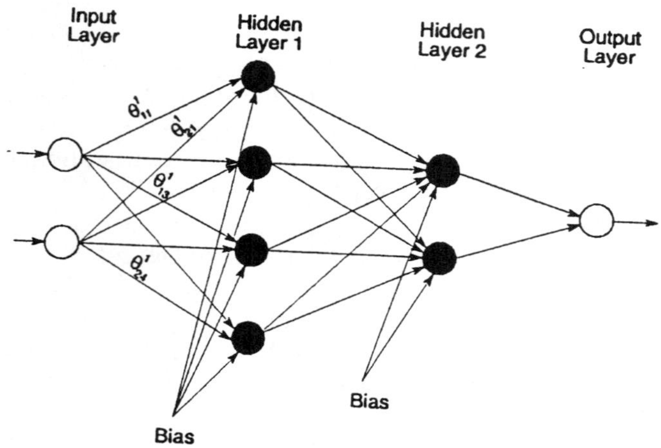

Fig. 12.1 NN (2,4,2,1).

(parameters) between node l in the input layer and node j in the first layer at a given iteration.

In this study we use the SPSA-based NN controller (Spall and Cristion, 1992) to estimate the NN weights vector to provide a set of control values that in turn drive the observation values close to the target value. The following subsection provides a brief review of the approach.

12.2.2 A brief review of the SPSA-based NN controller

The SPSA-based NN controller assumes that the number of layers and nodes of each layer are known and it only tries to estimate the parameters (weights). Consider a system output vector at time $k+1$ given by:

$$x_{k+1} = \phi(x_k, x_{k-1}, \ldots, x_{k-s}, u_k, w_k), s \geq 0 \qquad (12.3)$$

where $\phi_k(.)$ is generally unknown. The goal is to choose a sequence of control vectors $\{u_k\}$ in a manner such that the system output $\{x_{k+1}\}$ is close to the sequence of target vectors $\{t_{k+1}\}$, where the closeness is relative to the magnitude of the noise $\{w_k\}$ and the cost associated with the control. Assume that the goal is to find the control that minimizes the one-step-ahead loss function L_k, where the loss function is a function of x_{k+1}, t_{k+1} and u_k. A special case of the loss function is when the optimal controller value minimizes the square of the deviation from the target value.

In general, a vector of connection weights (Θ_k) of an NN is estimated considering a stochastic approximation algorithm of the form:

$$\hat{\Theta}_k = \hat{\Theta}_{k-1} - a_k \hat{g}_k(\hat{\Theta}_{k-1}) \tag{12.4}$$

where $\hat{\Theta}_k$ denotes the estimate of Θ_k at the given iteration; a_k is a scalar gain sequence satisfying certain conditions; $\hat{g}_k(.)$ is an estimation of the gradient of the loss function.

Since the function $\phi_k(.)$ is unknown, one cannot use the gradient descent updating rule which is commonly used in the literature. Spall and Cristion suggest that in such a case one uses an SPSA of $g_k(.)$ in estimation of the NN weights. The hth component of the gradient estimate, $\hat{g}_{kh}(.), h = 1, 2, \ldots, p$, via SPSA, is given by

$$\hat{g}_{kh}(\hat{\Theta}_{k-1}) = \frac{\hat{L}_k^{(+)} - \hat{L}_k^{(-)}}{2c_k \Delta_{kh}}, \quad \text{for } h = 1, 2, \ldots, p \tag{12.5}$$

where the notation is as follows.

- p represents the number of connection weights of NN.
- $\Delta_k = \{\Delta_{kh}\}$ is a sequence of p independent, identically symmetrically distributed (about 0) random variables with $E(\Delta_{kh})^{-2}$ uniformly bounded. Note that the above condition is a general condition (Δ_{kh} may not be either uniformly or normally distributed).
- $\{c_k\}$ is a sequence of positive numbers satisfying certain regularity conditions.
- $L_k^{(\pm)}$ are loss values based on $t_k + 1$, $x^{(\pm)}{}_k$, $u^{(\pm)}{}_k$.
- $u^{(\pm)}{}_k$ are control values and they are NN outputs. Their values are based on the NN p-element weight vector $\hat{\Theta}_k = \hat{\Theta}_{k-1} \pm c_k \Delta_k, t_{k+1}$, and $x_k^{(\pm)}$.
- $x_{k=1}^{(\pm)}$ are process outputs based on $u_k^{(\pm)}$.

To train the NN controller to produce an input vector u_k such that input vector drives the plant to the desired target state t_{k+1}, the following steps are recommended for the *non-resetting* method which will be employed in this study. For a review of the resetting method see Spall and Cristion (1992).

NN controller training steps: non-resetting method

1. Set $k = 0$.
2. Start with p-element weight vector Θ_k.
3. Generate $\{\Delta_{kh}\}$ according to the condition above and select a sequence $\{c_k\}$ that satisfies the regularity condition in Spall (1992).
4. Select $s(.)$ function. The form of $s(.)$ is commonly selected as $[1 + \exp(-I_j^i(k))]^{-1}, [1 - \exp(-2I_j^i(k))]/[1 + \exp(-2I_j^i(k))]$;

or $[\exp I_j^i(k) - \exp(-I_j^i(k))]/[\exp I_j^i(k) + \exp(-I_j^i(k))]$, where $I_j^i(k)$ is the input to the jth node in the ith layer.
5. Based on x_k, t_{k+1}, and p-element NN weight vector $\hat{\Theta}_k = \hat{\Theta}_{k-1} + c_k\Delta_k$, determine $u_k^{(+)}$ (outputs of the end layer's nodes) and generate $x_{k+1}^{(+)}$, using the relationship given in (12.3), and $\hat{L}_k^{(+)}$.
6. Given the system at $x_{k+1}^{(+)}$, determine $u_k^{(-)}$ based on $x_{k+1}^{(+)}, t_{k+1}$ and weight vector $\hat{\Theta}_k = \hat{\Theta}_{k-1} - c_k\Delta_k$. Then generate $x_{k+1}^{(-)}$ and $\hat{L}^{(+)_k}$.
7. Use algorithms (12.4) and (12.5) to generate $\hat{\Theta}_k$. Then, based on $\hat{\Theta}_k, x_{k+1}^{(-)}$ and t_{k+1}, generate u_k and corresponding x_{k+1}.
8. Set $k = k + 1$ and return to step 2.

Repeat the above steps until the weights yield a controller with a sufficient accuracy.

Spall and Cristion recommended SA gains of the form $a_k = a/k^\alpha$ and $c_k = c/k^\gamma$, where α and γ take values between 0 and 1, c and a are the initial values for learning rates. They also mentioned that the performance of the controller is sensitive to the initial value of the learning rate a and recommended that one has to start with a small value of a and increase the value very gradually for better results.

12.2.3 Statistical process control charts

Statistical control charts are widely used in industry for monitoring the quality of manufactured products. These schemes provide a set of criteria that enable one to judge, at any given time, whether the process generates the observations within an acceptable variation range. Once the control scheme is applied, it will sooner or later trigger an 'out of control signal', either because the underlying parameter (sometimes structure) has changed or simply because of the randomness inherently present in data, in which case the signal is referred to as a false alarm. In practice a control chart which is sensitive with respect to undesirable patterns of incoming observations and has a reasonable degree of protection against a false alarm is desired (Yashchin, 1993).

When dealing with autocorrelated data, the traditional Shewhart control charts are not appropriate tools for monitoring the process (Montgomery and Mastrangelo, 1991; Wardell, Moskowitz and Planta, 1992). The basis of existing efforts in constructing statistical control charts for autocorrelated data is to fit an appropriate time series model to the observations and then apply a control chart to the sequence of residuals from this model. The belief is that if the model is well-fitted, the residuals will behave like independent and identically random variables. Therefore, when control charts are applied to the sequence of residuals, if a shift occurs in mean or if variability increases, the identified model will no longer be correct and this model misspecification will be transferred to the residuals. Ultimately, the shift will be detected on the control chart applied to the residuals.

This study uses two statistical control charts which are appropriate for autocorrelated data [Alwan and Roberts, 1988; Montgomery, 1992; Yashchin, 1993). The two charts are the special cause control (SCC) chart and the exponentially weighted moving average (EWMA) chart.

The SCC chart is used to monitor a process after the process is modeled (for example by an autoregressive moving average model). The SCC chart is merely a run chart, but rather than plotting the actual observations, the residuals or one-step-ahead forecast errors are plotted. This chart is based on the assumption that the residuals are random, and are independent and identically distributed around 'zero' with small variance, when a well-defined model represents the process and appropriate control action is applied. However, if there are any disturbances to the process, the residuals will begin to show a departure from statistical control. Since it is common to assume that the mean of residuals is 'zero', the centerline of the SCC chart is 'zero'; similarly the standard deviation used in this case is the standard deviation of the residuals, not the standard deviation of observation.

The EWMA chart is designed to detect small shifts in the mean value by giving exponential weight to past observations, the more recent data receiving the heavier weights. Montgomery (1992) suggests that, in general, if the observations from the process are positively autocorrelated and the process mean does not shift too quickly, the EWMA with an appropriate value for smoothing constant will provide an excellent one-step-ahead predictor. The EWMA at time k is defined as:

$$Z_{k+1} = \lambda x_k + (1 - \lambda) Z_k, \quad 0 < \lambda \leq 1 \tag{12.6}$$

where λ is a smoothing constant which determines the weight given to past observations; x_k is the observation at time k; and Z_{k+1} is the smoothed value of observation based on the EWMA procedure given in (12.6). The control chart plots EWMA values on a chart with the following limits:

$$\begin{aligned} UCL_{k+1} &= Z_k + U_{\alpha/2}\sigma_p \\ LCL_{k+1} &= Z_k - U_{\alpha/2}\sigma_p \end{aligned}$$

where σ_p is the standard deviation of one-step-ahead errors; UCL and LCL are the upper and lower control limits, respectively; and $U_{\alpha/2}$ is the upper $\alpha/2$ percentage point of the unit normal distribution.

When data is autocorrelated, it is suggested that one searches for the value of λ that minimizes the sum of the squares of one-step-ahead prediction errors, and if λ is estimated over a record of n observations, then one divides the sum of the squared prediction errors by n to produce an estimate of σ_p^2 (Montgomery, 1992).

The criterion for measuring the performance of the above charts is the average number of observations taken before a signal triggers the existence of an assignable cause.

12.3 SIMULATION EXPERIMENTS

In this study the batch polymerization example provided by Vander Wiel *et al.* (1992) is employed and it is assumed that the stochastic model they constructed represents the true dynamics of the polymerization process, but that the knowledge about the process dynamics does not exist. Then the non-resetting SPSA-based NN controller is used to simulate control values that reduce the polymer viscosity variations about the target value. The SCC chart is used for monitoring the viscosity variations. After the NN controller provides the output in the tolerance limit, a change will be initiated in the process dynamics and both charts will be employed to monitor the polymerization process. Then, the speed of the SCC and EWMA charts will be examined in triggering signals for existence of an assignable cause. The study will also report the speed of NN controller in producing control values that in turn drive the viscosity variations to within the acceptable variation range.

12.3.1 Simulation study

For the batch polymerization example provided by Vander Wiel *et al.* (1992), it is assumed that each batch cycle consists of the following steps: charge the reactor, run the reaction, empty the reactor contents into a holding tank for subsequent processing, engage cleaning procedures and recharge to begin the next batch. Intrinsic viscosity, a key quality characteristic of the polymer, is measured at the completion of each batch. Turnaround time is such that the viscosity measurement from the most recent batch produced in a given reactor is available when the reactor is prepared for a new batch. The statistically significant viscosity variations about a target level of 100 viscosity units is observed.

In analyzing the cause of viscosity variations, Vander Wiel *et al.* reported that autocorrelation among viscosity measurements exists. Autocorrelations stem from several factors. For example, mechanical considerations prevent a reactor from being fully emptied between batches, or economic and environmental considerations necessitate the recovery and reuse of certain reaction components (e.g. solvents). They also reported that the level of catalyst has the highest effect on the level of viscosity, and must be adjusted in order to reduce the viscosity deviations from the target value. Based on the observed intrinsic viscosity deviations from its target value and the catalyst deviations from its nominal value, Vander Wiel *et al.* provided the following empirical model for the batch k.

$$x_k = 0.8x_{k-1} + 1.5u_{k-1} - 1.2u_{k-2} + w_k - 0.22w_{k-1} \qquad (12.7)$$

where x_k is observed viscosity deviation from a 100 unit target viscosity, u_{k-1} is the catalyst deviation from nominal (50 units), w_k is roughly independent of $N(0, \sigma_w^2)$, and $\sigma_w = 2.798$. Equation (12.7) employs the compromise parameter estimate values of Vander Wiel *et al.* (1992) empirical study. Under the assumption that the most recent viscosity measurement is available, and based on the minimum mean square errors, Vander Wiel *et al.* reported the following so-called pure-one-step adjustment rule:

$$u_{k-1} = 0.8u_{k-2} - 0.4x_{k-1} \qquad (12.8)$$

This simulation study employs the non-resetting SPSA-based NN controller discussed in section 12.2 instead of algorithm (12.8). The data will be generated according to the empirical model (12.7) and the controller here does not assume knowledge of (12.7). For a reactor, the SPSA-based NN controller adjusts the level of catalyst for the ith batch based on the viscosity deviation of the $(i-1)$th batch from the target value, and the catalyst deviation of the $(i-2)$th batch from the nominal value.

The NN of this study has an input layer with three nodes (representing t_{k+1}, x_k, u_{k-1}), and an output layer with one node (representing u_k) and the two hidden layers with 20 and 10 nodes, respectively, as in the Spall and Cristion study. The initial weights are generated from a uniform distribution $(-0.1, 0.1)$. Since the process is stationary, decaying SA gains of the form $a_k = a/k^{0.602}$ and $c_k = c/k^{0.101}$ with $c = 0.48$ and $a = 0.01$ are used. The values of x_1, u_0, w_0 are set to 20.0, 4.0, and 0.0 respectively. Since x_k represents the kth batch viscosity deviation from the target value, x_1 is assigned the value 20, a value more than $3\hat{\sigma}_w(8.394)$, and t_k is assigned the value zero throughout the simulation study. The values of u_0, w_0 are arbitrarily assigned. These values will affect the first few values of the simulation runs but they will not affect the trend of performance of the controller. The hidden layer nodes are hyperbolic tangent function (e.g. $(e^I - e^{-I})/(e^I + e^{-I})$ for input I) while the output node is a linear function (simply output = input). The perturbations $\{\Delta_{kh}$ are generated from a Bernoulli ± 1 distribution which satisfies the required conditions. Each node takes as an input (I) the weight sum of outputs of all nodes in the pervious layer plus a bias weight not connected to the rest of the network.

The SCC chart with control limits of $\pm 3\hat{\sigma}_w$ is used for monitoring the viscosity variations. The EWMA cannot be used at this stage of the study since the value of its parameters (λ and σ_p^2) cannot be estimated. The NN controller has no knowledge about the dynamics of the process and it learns about the process dynamics through the first iterations. Therefore first viscosity deviations (errors) are not independent and their values should fall outside the SCC control limits. It is appropriate that, at this stage, we use the SCC chart to study the learning speed of the NN controller in readjusting the levels of catalyst and producing polymers with viscosity values close to the target value.

This study conducts five independent simulation runs, each of 1000 iterations. Each simulation run represents the performance of one reactor. The result of one iteration of a simulation run represents the polymer viscosity deviation of one production batch from the target value. For every simulated reactor and every sequence of 100 simulated batches, the study calculates the number of simulated production batches whose viscosity deviations from the target value fall outside the SCC control chart limits. Figure 12.2 presents the average performance of the five simulated reactors. As Fig. 12.2 indicates, the percentage of batches whose viscosity deviations fall outside the SCC control chart limits decreases over time and the rate of reduction is relatively fast.

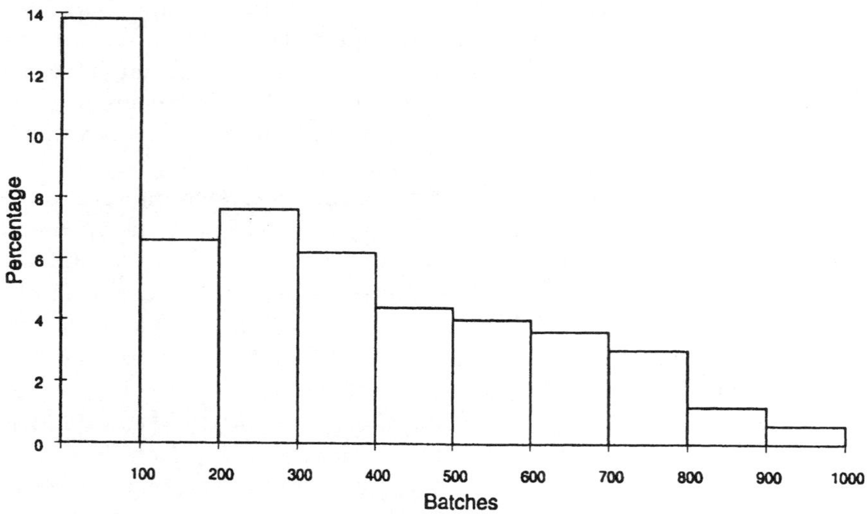

Fig. 12.2 Percentage of simulated viscosity deviation values that are outside the SCC chart limits.

12.3.2 Examining the performance of the SCC and EWMA charts

In this subsection it is assumed that after producing an additional 399 simulated batches the process dynamics of polymerization, for all five reactors, changes from (12.7) to:

$$x_k = 0.8x_{k-1} + 0.25u_{k-1}x_{k-1} + 1.5u_{k-1} - 1.2u_{k-2} + w_k - 0.22w_{k-1}.$$

Both SCC and EWMA charts are used to monitor the viscosity variations. Concurrently the speed of the SCC and EWMA charts in signaling the change in the process structure will be studied.

The assumption of sudden change in the underlying structure of the process is determined completely subjectively. It represents the fact that there is a possibility of change in the underlying nature of an operation process, for example, due to change of raw materials. The change calls for readjusting the catalyst levels to produce products with viscosity levels close to the target viscosity value.

For this experiment, assumptions and the SPSA-based NN controller structure are the same as those used in section 12.3.1. But SA gains are constant ($a = 0.0007$ and $c = 0.46$) instead of having a decaying nature. The constant rates are recommended by Spall and Cristion for systems with time-varying dynamics.

The EWMA chart and the SCC chart are developed as described in section 12.2.3. The data of the first 399 simulated batches of each reactor is used to estimate the parameters of the EWMA chart (λ and $\hat{\sigma}_p$) for that reactor's performance. For both charts, the same level of significance is used (i. e. $U_{\alpha/2} = 3$ is used for the EWMA control limits). The criterion for evaluating the performance of a control chart is the average number of simulated batches required before a chart detects a shift in the process dynamics, once the shift has occurred. The occurrence of three or more consecutive viscosity deviation values outside the control limits of a chart will be labeled as a signal informing the existence of a special cause.

For this experiment, we also conducted five independent simulation runs, each of 1000 iterations. The results of the simulation runs indicated that on average the SCC chart provided signals after 17.25 simulated production batches; whereas, the EWMA could not trigger a signal for two out of five independent simulated reactors. As a result the average number of batches before the trigger happens could not be calculated. Accordingly, we conducted the EWMA chart with $U_{\alpha/2} = 1.96$. On average, the EWMA chart with $U_{\alpha/2} = 1.96$ provided signals after 18.75 simulated production batches.

For studying the average speed of the SPSA-based NN controller after the change has been triggered, the EWMA chart with $U_{\alpha/2} = 1.96$ along with the SCC chart are used. Figures 12.3 and 12.4 present the average speed of the SPSA-based NN controller. As the figures show, the NN controller, when coupled with the SCC chart, has provided a relatively faster learning rate than when it was coupled with the EWMA chart. As Fig. 12.3 shows, the percentage of the viscosity deviation values that has fallen outside the SCC control chart limits has dropped substantially over time, and the SCC chart has provided fewer false alarms (fewer viscosity deviation values outside the control limits for the first 399 iterations) than the EWMA chart has.

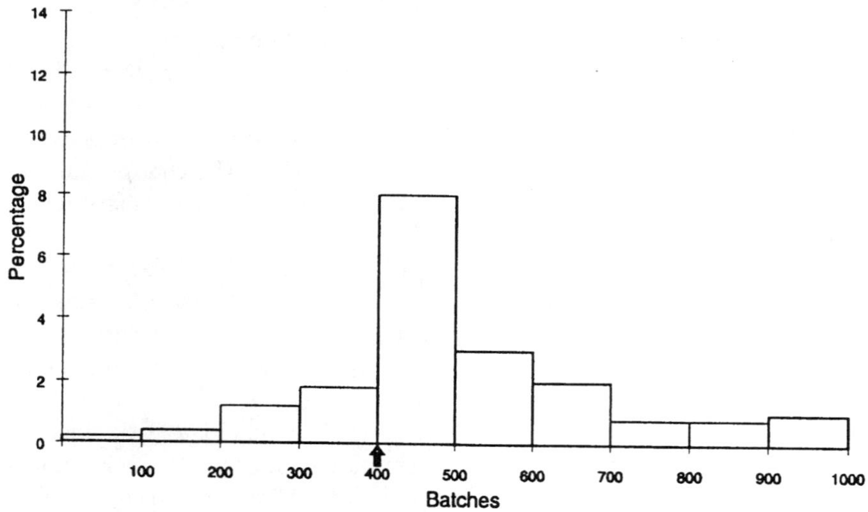

Fig. 12.3 Percentage of simulated viscosity deviation values that are outside the SCC chart limits when the process dynamics have changed.

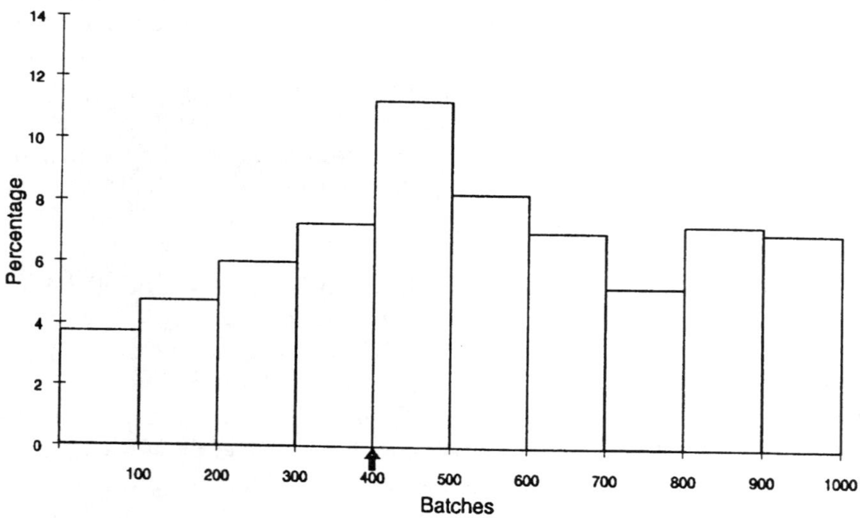

Fig. 12.4 Percentage of simulated viscosity deviation values that are outside the EWMA chart limits when the process dynamics have changed.

On the whole the SCC chart, when coupled with the NN controller, performed better than EWMA chart, in that it provided a faster signal for an assignable cause and also provided fewer false alarms. The poor performance of the EWMA chart could be related to the sudden change that we caused in the process dynamics, and the fact that the EWMA chart is intended for cases in which the processes' structures do not change too quickly.

12.4 CONCLUSIONS

This study employed the SPSA-based NN controller, an engineering process control approach, to reduce the variation in viscosity values of a polymer around the target value and assumed no knowledge about the process dynamics. A sudden change was initiated in the underlying structure of the process; then the performance the SCC and EWMA charts were examined in signaling the change of the process structure. The speed of the NN controller was also studied in responding to the change in process structure and in readjusting the levels of catalyst.

The simulation results indicated that the learning speed of the SPSA-based NN controller about the process dynamics is relatively fast. The findings of this study also indicated that the SCC control chart performed better than the EWMA chart based on either of the following criteria:

1. the least number of false signals,
2. the average number of simulated batches required before a chart indicates that there is a shift in the process dynamics, once the shift has occurred.

It should be mentioned that the Spall and Cristion approach assumes that an adequate structure of the NN is known. In fact the determination of the structure of the NN by itself is important. This simulation study employed the same number of hidden layers and nodes for each hidden layer as the Spall and Cristion studies, and has obtained satisfactory results. The findings of this study support Spall and Cristion's statement that the performance of the SPSA-based NN controller tends not to be very sensitive to the number of layers or nodes in a layer of an NN. However, a better structure may have resulted in a faster convergence.

On the whole, the findings of this simulation study suggest that the SPSA-based NN controller seems a promising approach for improving the quality of output of an unknown or a complex nonlinear process. The results of this study also imply that the SCC control chart is a better monitoring tool than the EWMA chart for autocorrelated data when the process dynamics are unknown and an NN controller is employed.

ACKNOWLEDGEMENTS

Part of this research was supported by the California State University–Dominguez Hills RSCAAP Summer Fellowship.

REFERENCES

Alwan, L.C. and Roberts, H.V. (1988) Time series modeling for statistical process control *Journal of Business and Economic Statistics*, **6**, 87–95.

Box, G. (1993) Quality improvement – the new industrial resolution. *International Statistical Review*, **61**, 3–19.

Box, G. and Draper, N. (1987) *Empirical Model-Building and Response Surfaces*, John Wiley and Sons, New York, pp. 10–12.

Chryssolouris, G., Lee, M. and Domroese, M. (1991) The use of neural networks in determining operational policies for manufacturing systems. *Journal of Manufacturing Systems*, **10**, 166–175.

Faltin, F.W., Hahn, G.J., Tucker, W.T. and Vander Wiel, S.A. (1993) Algorithmic statistical process control: some practical observations, *International Statistical Review*, **61**, pp. 67–80.

Gupta, P.Y. and Kumar, S. (1991) Controlling the production process through statistical process control. *Manufacturing Review*, **4**, 18–31.

Montgomery, D.C. (1992) The use of statistical process control and design of experiments in product and process improvement. *IIE transactions*, November, 4–17.

Montgomery, D.C. and Mastrangelo, C.M. (1991) Some statistical process control methods for autocorrelated data. *Journal of Quality Technology*, **23**, 179–204.

Narendra, K.S. and Parthasarathy, K. (1990) Identification and control of dynamical systems using neural network, *IEEE Transactions on Neural Networks*, **1**, 4–27.

Nguyen, D.H. and Widrow, B. (1990) Neural networks for self-learning control systems. *IEEE Control Systems Magazine*, **10**, pp. 18–23.

Smith, A.E. and Dagli, C.H. (1991) Controlling industrial process through supervised, feedforward neural networks. *Computers and Industrial Engineering*, **21**, 247–51.

Spall, J.C. (1992) Multivariate stochastic approximation using a simultaneous perturbation gradient approximation. *IEEE Transactions on Automatic Control*, **37**, 332–41.

Spall, J.C. and Cristion, J.A. (1992) Direct Adaptive control of nonlinear systems using neural networks and stochastic approximation, in *Proceedings of the IEEE Conference on Decision and Control*, Tucson, AZ, pp. 878–83.

Spall, J.C. and Criston, J.A. (1994) Nonlinear adaptive control using neural networks: estimation with a smoothed simultaneous perturbation gradient approximation. *Statistica Sinica*, **4**, 1–27.

Tulunay, E. (1991) Introduction to neural network and their application to process control, in *Neural Networks: Advances and Applications* (Ed. E Gelenbe), Elsevier, Amsterdam, pp. 241–73.

Udo, G.J., (1992) Neural networks applications in manufacturing processes. *Computers and Industrial Engineering*, **23**, 97–100.

Vander Wiel, S.A., Tucker, W.T., Faltin F.W. and Doganaksoy, N. (1992) Algorithmic statistical process control: concepts and an application. *Technometrics*, **34**, 286–97.

Wardell, D.G., Moskowitz, H. and Plante, R.D. (1992) Control charts in the presence of data correlation. *Management Science*, **38**, 1084–105.
Yashchin, E. (1993) Statistical control schemes: methods, applications and generalizations. *International Statistical Review*, **61**, 41–66.

CHAPTER THIRTEEN

Constraint-based genetic algorithms for concurrent engineering

Denis R. Cormier and Peter J. O'Grady

13.1 INTRODUCTION

Concurrent engineering is the consideration of the factors associated with the life-cycle of a product during the design phase. These factors include product functionality, manufacturing, assembly, testing, maintenance, reliability, cost and quality (O'Grady and Young, 1991). Concurrent engineering is important because it is at the design stage that such aspects as product quality and cost are specified. The degree of implementation of concurrent engineering can be thought of as being divided into the following four categories (O'Grady and Young, 1991).

1. One-way: sometimes called 'over the wall: designs are developed by designers in relative isolation from other functions. These designs are then passed to manufacturing. This is, in many ways, the traditional mode of operation for many companies, with little attention being paid to improving the design from the point of view of the various life-cycle factors.
2. Stunted two-way: designs are evaluated, and altered where necessary, after the designs are complete.
3. Two-way: designs are evaluated at stages during the design process.
4. Integrated: designs are continuously evaluated.

A full concurrent engineering implementation would involve category 4 above, i.e., *integrated* design incorporating life-cycle factors. In such a concurrent engineering implementation, designs are continuously evaluated from the first conceptual design all the way through to the final engineering drawings. By considering and integrating the life-cycle factors early in the design stages, overall designs can be improved (Bedworth, Henderson and Wolfe, 1991). One indication of the potential value of a concurrent engineering system can be seen by comparing the cost of product changes made during the design phase to the cost of changes postponed until the product reaches manufacturing. Once a product design is finalized, it is relatively costly to identify and implement modifications. It has been reported that upwards of 70% of a product's manufacturing cost is dictated by decisions made during the product design stage (Andreasen, Khler and Lund, 1983). In addition, other benefits such as improved quality and reduced lead-times are to be expected. Thus, employing concurrent engineering systems to improve a design offers the potential of reducing costs, raising quality and improving service to the customer.

While there is the potential for substantial benefits obtained from implementing concurrent engineering, the implementation can be an onerous and difficult task due to the number of considerations that have to be brought to bear during the design activity. The implementation can be aided using concurrent engineering systems. These are systems, both manual and computerized, that can assist in carrying out the concurrent engineering task. The emphasis in the development of concurrent engineering systems is on coordination to eliminate problems due to lack of communication and compatibility among different areas of concern. The primary capability of concurrent engineering systems involves handling different, often conflicting goals, that need to be harmonized and integrated into a working whole by taking many constraints restricting the design into account. A key element of this is the ability to determine improvements that can be made to a design.

This chapter addresses this issue. It is concerned with devising an approach that can determine an improved design. This approach is based on new algorithms that extend the conventional genetic algorithm to handle mixtures of algebraic, conditional and database constraints with quantitative, symbolic and qualitative variables. This approach we call a constraint-based genetic algorithm (CBGA). The chapter describes the basis for CBGA and describes an example design session using CBGA. The format of this chapter is as follows. An overview of approaches to concurrent engineering is first given, and this is followed by a discussion of the research issues associated with implementing support systems for concurrent engineering. Genetic algorithms and constraint network modeling are then described. An overview of the CBGA system is then given together with a sample design session using the CBGA.

13.2 APPROACHES TO CONCURRENT ENGINEERING SYSTEMS

There are a number of techniques that can be used in concurrent engineering systems to advise designers on aspects that reduce life-cycle problems. These include the use of design teams, design handbooks, checklists and structured procedures, manufacturing simulation and process planning, rule-based expert systems, and constraint networks.

13.2.1 Design teams

One way of integrating design and life-cycle functions is by the use of design teams. This approach requires early and frequent interaction between design and the other life-cycle functions. It has traditionally been accomplished with post-design reviews which are usually considered to be less effective than reviewing the design while it is being developed (Heller, 1971). A better approach is to have team members actually design the product and process together (Maddux and Jain, 1986; Evans, 1988). Group approaches provide alternative perspectives to the design problem, and group approaches can also be useful for initial idea generation. However, group decision making (for creative tasks, in particular) is difficult (Harfmann, 1987). The operation of design teams would be further improved by the use of suitable techniques and systems that advise on good designs from the life-cycle perspective.

2.2 Handbooks

The most conventional approach to advising on design is to provide compiled life-cycle experience in handbooks, standards, and design guides. The designer is then responsible for considering and integrating this documented information into design decisions (Bralla, 1986; Trucks, 1987; Stein and Strasser, 1986). However, this approach has not been entirely satisfactory since the designers still have to cope with the high volume, variety and interdependence associated with life-cycle knowledge.

13.2.3 Checklists and procedures

Post-design review checklists and procedures are available for some aspects of life-cycle design, notably manufacturing. The designer typically classifies the design and obtains guidelines for that class of design feature or problem (Eversheim and Muller, 1984; Heller, 1971; Oakley, 1984; Redford, Swift and Howie, 1981; Redford, 1986; Boothroyd and Dewhurst, 1988). The main advantage of classification tables and checklists are that they are relatively easy to construct and can be used in a straightforward manner. However, these approaches tend to depend on accurate cost information. This is considered by some to be a major weakness (Kroll, Lenz and Wolfberg, 1988) since accurate cost information is often difficult to obtain (Heller, 1971). Others would suggest that the biggest danger in using this type of evaluation is that it will be approached in a mechanical way and the results will be viewed with greater accuracy than they actually merit (Oakley, 1984). Haeusler (1981) points out that structured procedures can conceal the danger of sub optimization. A further potential problem is that the rigid structuring may also be unrealistic (Kroll Lenz and Wolfberg, 1988; Dieter, 1983; Lu, 1986; Gross *et al.*, Rosen, Riley and Erdman 1987; Heller, 1971).

13.2.4 Process planning and manufacturing simulation

The use of process planning and manufacturing simulation systems (such as NC tool cutting graphics systems) to aid the designer in considering the life cycle aspect of manufacturing has been proposed by several researchers (Maddux and Jain, 1986). Such approaches suffer from the disadvantage that they tend to concentrate only on process operations and provide only indirect feedback to the designer on potential problems.

13.2.5 Rule-based expert systems

There are relatively few computerized systems that support concurrent engineering. Those that do exist are mainly rule-based expert systems (Chao, 1985; Sevenler *et al.*, 1986). Work on design for assembly has produced several rule-based systems, including (O'Grady and Oh, 1990, 1991; Oh, O'Grady and Young, 1995) the following.

- A Systematic Approach to Design for Assembly from Lucas Engineering & Systems Ltd. (Hollingum, 1989). This provides sequential iterative analysis of function, feeding, gripping, insertion and assessment, and gives out cost-related indices. It has a similar aim to that of the Boothroyd and Dewhurst (1988) system in that it strives for reductions in the number of components. A single system is used for both robotic and automatic assembly.
- ADAM (Assisted Design for Assembly and Manufacture) provides verbal advice and parts list on minimizing parts, rationalization and insertion guidelines (Sackett and Holbrook, 1988).
- DACON (Design for Assembly Consultation) attempts a CAD interface. It provides redrawing of parts as output from an analysis of the design by an expert system (Swift, 1987).
- PACIES (Part Code Identification Expert System) (Chen and Young, 1988) generates three digit U-Mass part codes based on the shape, important features and symmetries of the part. The system is implemented under DOS using the micro-PROLOG programming language, and consists of a rule set containing 163 rules.
- A prototype expert system for the design for assembly of surface-mounted printed circuit boards has been developed by Ball (1988).

The advantages of rule-based systems are that they are relatively simple. There are, however, several disadvantages of rule-based systems. Most are applicable to a very small application area and have problems in being expanded to handle a wider application area. They are unable to allow designers to approach the design problem from whatever perspective is appropriate given the information available at a particular stage of the design process. Furthermore, a prime requirement of supporting concurrent engineering is the ability to represent the mutually constraining influences that different aspects of a product's life-cycle exert on each other. This requirement is not easily achieved using rule-based systems.

13.2.6 Constraint networks

It is generally agreed that design activities focus on the recognition, formulation and satisfaction of constraints (Gross *et al.*, 1987; Wu, 1988). Based on this notion, the constraint network methodology has been developed to simulate the way designers think about and attack problems. Constraint networks are a collection of constraints, interconnected by virtue of shared variables relating to the network. These networks offer a logical means of finding feasible solutions to complex design and planning problems by representing dependence relationships among the variables. When implemented in a concurrent engineering system, a tool is created for interactive, user-involved design and planning. A number of constraint network systems have been developed, including SPARK and Saturn (Young, Greef and O'Grady, 1992).

Since the pioneering work of Sutherland's (1963) Sketchpad, a constraint-based system for drawing and editing pictures on a computer, there have been a number of constraint programming languages, such as CONSTRAINTS (Sussman and Steele, 1980), ThingLab (Borning, 1979), IDEAL (Van Wyk, 1981), MOLGEN (Stefik, 1981a, b), TK!Solver (Konopasek and Jayaraman, 1984), Juno (Nelson, 1985), CLP($\Re$) (Heintze, Michaylov and Stuckey, 1987), CHIP (Van Hentenryck, 1989) and Bertrand (Leler, 1988). Several existing constraint languages are reviewed in Leler (1988) Most of them are heavily dependent on a specific domain, such as graphical user interface design, electric circuit design, physical simulations, molecular genetic experimental design, puzzle problems and mechanical design and analysis. Consequently, their capability to handle constraints is limited to certain classes that are relevant to the domain problems, such as numeric, linear equations and inequalities. Some general systems, for example CHIP and CLP($\Re$), have tried to expand the constraint handling capability. SPARK was one of the earliest systems specifically oriented towards concurrent engineering systems. SPARK is briefly reviewed below while its successor Saturn is described later in this chapter, together with a deeper description of constraint networks applied to concurrent engineering systems.

SPARK is a constraint network system with frame-based inheritance and was one of the earliest constraint network systems developed (Young, Greef and O'Grady, 1992). The language facilitates building sets of constraint networks which collectively take on the form of design advisory systems. Constraint networks are interconnected through shared variables, and the user's objective is to find a set of variable values that does not violate any of the constraints. Variable values are either determined automatically by the system or supplied by the user. Once a value is established, it is propagated bi-directionally through the system. In using SPARK to develop a constraint network, the designer or process planner is utilized as an *a priori* knowledge source when the inference mechanism is developed (Kim, O'Grady and Young, 1991, 1992, Kim *et al.*, 1992b).

Constraint networks are a fairly recent development. However, they have been applied to an extensive range of design and concurrent, engineering problems across a diverse range of industries (Kim, O'Grady and Young, 1991, 1992a; Kim *et al.*, 1992b; Liau and Young, 1992; Oh, O'Grady and Young, 1995; O'Grady, Young and Smith, 1991; O'Grady, Kim and Young, 1994; O'Grady and Young, 1991).

13.3 RESEARCH AREAS IN CONCURRENT ENGINEERING SYSTEMS

The development of a system to support concurrent engineering is a challenging task. The main requirements for the concurrent engineering system can be placed into the following three categories (O'Grady and Young, 1991).

1. **Interface to CAD systems**: the concurrent engineering system should readily interface to CAD systems.
2. **Domain knowledge**: the system should have the requisite knowledge needed to evaluate a design and determine improvements to the design.
3. **Logic and internal operations**: the concurrent engineering system should be flexible enough to allow the design problem to be approached from a variety of viewpoints; it should allow the designer to design despite the absence of complete information; and it should handle the large volume, variety and interdependence of life-cycle information. The central aim is that the concurrent engineering system should assess the design and determine improvements that can be made.

This chapter is concerned primarily with category 3 above – *logic and internal operations*. In particular it is concerned with the central aim of determining improvements that can be made to the design. The methodologies for determining improvements can be grouped as follows (O'Grady, Young and Smith, 1991).

Design rating involves evaluating the whole design and giving a rating as to its acceptability. The rating would usually be numeric. Design rating approaches have been used mainly for evaluating the ease of manufacture or assembly of a product (Bao, 1988; Zorowski, 1986). In these approaches an overall numeric rating is given to the design with little or no advice being passed to the designer on how to improve the design. A completed design is usually evaluated and this may be less effective than evaluating a design as it is being completed. A further disadvantage of design rating systems is that the domain considered is usually fairly narrow.

Structured approaches rely on either checklists or procedures to carry out a post-design review. Structured approaches, including post-design review checklists and procedures, are available for some aspects of the life-cycle, notably manufacturing. The designer typically classifies the design and obtains guidelines for that class of design feature or problem (Eversheim and Muler, 1984; Heller, 1971; Oakley, 1984; Redford, Swift and Howie, 1981; Boothroyd and Dewhurst, 1988). Major problems with these structured approaches are the following.

1. They often depend on accurate cost information and this may be difficult to obtain (Heller, 1971; Kroll Lenz and Wolfberg, 1988).
2. The results may be viewed with greater accuracy than is actually warranted (Oakley, 1984).
3. The rigid structuring may be unrealistic (Kroll, Lenz and Wolfberg, 1988; Dieter, 1983; Rosen, Riley and Erdman, 1987; Heller, 1971).
4. They may result in suboptimization (Haeusler, 1981).

Reference approaches rely on the designer accessing information on improving the design. The information would usually be contained in a reference book or database. Reference approaches include the use of handbooks, standards and design guides. The designer is responsible for considering and integrating this information into design decisions (Bralla, 1986; Trucks, 1987; Stein and Strasser, 1986). The major disadvantage of reference approaches is that the designer has to manage the high volume and variety of life-cycle information.

Design advice involves examining a design or design requirement and giving the designer advice about how to approach the design so as to best meet life-cycle requirements. Note that the designer can accept or reject the advice. This has been the subject of much recent work. While the design advice approach has much to commend it, the need for the user to make a large number of decisions can lead to a less than ideal design.

Automatic design involves examining the design requirements and automatically (or semi-automatically, i.e. with some, reduced, human input) producing a design that satisfies life-cycle requirements. The complexity of the design process, and the need to consider a wide variety of concurrent engineering views, means automatic design is a challenging task with many demanding research issues.

Ideally, a methodology for design improvement should put a designer's knowledge and experience to good use and should assist in the design optimization task. Design optimization is very difficult to do manually due to the need to simultaneously change many variable values in order to escape locally optimal designs. Furthermore, these complex design changes are heavily constrained due to the many life-cycle factors.

The CBGA system described in the remainder of this chapter was developed in response to the need for computerized support for design improvement. The CBGA combines various aspects of several design improvement methologies including the design advice and the semi-automatic design methodologies. In particular, the CBGA employs a constraint management system which interactively provides advice to designers as they work. Furthermore, it semi-automatically performs design optimization through the use of a constraint-based genetic algorithm. This approach is effective in that it puts designers' experience to good use and keeps them 'in the loop', yet frees them from the burden of trying to manually perform global optimization in such a complex environment.

13.4 CONSTRAINT NETWORK MODELING AND GENETIC ALGORITHMS

This section gives a description of constraint network modeling and genetic algorithms.

13.4.1 Constraint network modeling

Concurrent engineering problems can be declaratively modeled by constraint networks. Each piece of design knowledge, as well as restrictions constraining the design, are specified by a form of constraint in the network model. Thus, a concurrent engineering model is written by a conjunction of the constraints as follows:

$$a\_feasible\_design(T) :- \bigwedge_{i=1}^{m} \phi(\Gamma_i, T_i, A_i, F_i),$$

where T is the set of all variables in the model, and each ϕ_i is a constraint defined by Γ_i, T_i, A_i and F_i, which are sets of relation symbols, variables, constants and functional symbols, respectively. Obviously, $T = \bigcup_i T_i$. The purpose of such a model is to find a solution that satisfies every constraint $\phi_i(\Gamma_i, T_i, A_i, F_i)$.

A constraint specification represents a convenient form of expressing declarative knowledge, allowing designers to focus on local relationships among entities in the domain. A set of constraints collectively defines the feasible solution space where the modeled system can be set up. An assignment of values to all the variables, such that all the constraints are simultaneously satisfied, is called a solution. A constraint network problem is a problem of finding such a solution or all the solutions. Formal definitions of the constraint network problem are found in Dechter (1992) and Mackworth (1992).

Saturn is a constraint modeling system that can be used as part of a concurrent engineering system (Fohn *et al.*, 1993; 1994; Greef *et al.*, 1994). Saturn incorporates a constraint management system that implements theorem proving techniques based on resolution, relational database theory and non-monotonic deduction. The former two techniques operate on the constraint-based model to infer satisfied constraints while deducing attribute values that cannot violate other mutual constraints within the model. The latter technique supports a designer when retracting previous design decisions. Working in concert, these techniques provide an interactive constraint satisfaction environment capable of providing guidance and feedback to the designer during a design session. The key advances that Saturn brings to concurrent engineering systems are in the user interface, the interface to database systems, the provision of macros and in the dependency recording mechanism and inference mechanism. Some work has been done in examining the effects of imprecision (Kim *et al.*, 1995) and on the effects produced by large-scale concurrent engineering problems (Kim and O'Grady, 1996).

Saturn's constraint modeling system described thus far is very useful as an interactive design advice tool which allows users to quickly locate feasible design alternatives. However, it does not provide users with a mechanism to easily search for better solutions (i.e. design optimization).

13.4.2 Genetic algorithms

The genetic algorithm (GA) simulates evolution by imitating genetic inheritance and employing a survival of the fittest strategy. The algorithm is rather unusual in that it starts with a population of different solutions to the problem at hand (called the initializaion function). Each solution is evaluated according to the problem's performance metric(s) and only the better solutions tend to be retained (i.e. survival of the fittest designs) (called the evaluation function). Next, random portions of these surviving solutions are spliced together to form entirely new designs (called the crossover function). In essence, the resulting solution (child) inherits design attributes from its two contributing solutions (parents). In addition, a random mutation (the mutation function) can be carried out to produce new varieties of solutions. The process can proceed for many generations until good enough design alternatives are found.

Genetic algorithms are unique when compared with other optimization techniques in the sense that they operate on a population of many designs. As a result, this collection of design alternatives represents a diverse sampling of many portions of the search space. This parallel sampling is what allows the GA to effectively explore promising regions of the search space and to escape local optima. However, it is the evaluation–selection–recombination cycle that allows the GA to efficiently exploit knowledge gained during search. For extremely large problems, such as those encountered in concurrent engineering, this ability to balance exploration versus exploitation is what makes the GA a powerful search technique.

The GA has been used in a wide variety of manufacturing-related tasks including process planning (Vancza and Markus, 1991), scheduling (Lawton, 1992), (Husbands, Mill and Warrington, 1991), and layout planning (Tam, 1992), (Cohoon *et al.*, 1991). Each of these problems involve complex domains for which no clear-cut optimization approach has been found to be superior. DeJong (1988) states that 'it is precisely in these circumstances (large, complex, poorly understood search spaces) that one should consider exploiting the power of genetic algorithms.'

(a) Initialization

As previously stated, the GA operates on a population of solutions, not a single solution. Before the algorithm can be run, the initial population of solutions has to be created. The initialization function is responsible for assigning values to the n attributes of all m members of the population. For concurrent engineering problems, each of the n attributes is defined over a domain:

$$D = \{d_1, d_2, \ldots, d_n\},$$

where d_j can specify real numbers, integers, strings, database columns (sets) or any other arbitrary data type.

This initial population can include manually created designs, randomly created designs or designs generated through the application of heuristics. A combination of both randomly and non-randomly created designs appears to represent a compromise between the need for good exploration and the practical limits on the amount of time available to perform the search.

(b) Evaluation

In order for the GA to decide which designs are best, each design in the population needs to be evaluated according to some performance metric. Design metrics for concurrent engineering involve multiple objectives, where the objectives correspond to product functionality, manufacturability, ease of assembly, reparability and so on. Following each generation is an evaluation function where every member of the population is evaluated with respect to the multiple objectives. Given n objectives, the evaluation score of member i for objective l during generation k is denoted by:

$$\text{Eval}_{i,l,k} = f(a_{i,1,k}, a_{i,2,k}, \ldots, a_{i,n,k}, c_1, c_2, \ldots, c_z),$$

where $a_{i,j,k}$ denotes the value of the jth variable of population member i during the kth generation. If each member of the population is ranked from best to worst, based on each performance metric, then a simple weighted average of a member's ranks may be used to obtain a composite fitness measure. This rank-based approach avoids some of the difficulties associated with normalizing the various concurrent engineering performance metrics which may be expressed in different units or scales.

(c) Selection schemes

Based on the evaluation score for each alternative, the next step is to select the best designs for survival (i.e. survival of the fittest designs). Given a population of m members, only a subset consisting of the better members is stochastically chosen for survival. Each member of the population is ranked according to fitness scores. The two prevailing selection schemes currently used involve proportional selection and rank-based selection. Regardless of the approach used, design alternatives with the best evaluation scores have the highest probabilities of being selected.

With proportional selection, the probability of a design's survival is directly proportional to that design's fitness relative to the overall population fitness. Since these fitnesses sum to 1, each member's relative fitness represents the probability of survival. Survivors are selected by generating random numbers and comparing them with the cumulative distribution function of the population fitness.

One problem with the proportional selection scheme is its tendency to prematurely converge on a single solution. This problem stems from the fact that the presence of a single 'superfit' individual generally results in that individual's repeated selection while every other individual is weeded out. In response to this problem, rank-based selection schemes have been developed. Unlike proportional selection, rank-based selection ranks the population members from best to worst, based on evaluation scores, and then computes the probability of survival, based on the rank rather than the actual score (Whitely, 1989). For example, Michalewicz (1992) calculates the probability of survival using:

$$\text{Prob(rank)} = q(1-q)^{\text{rank}}$$

The relative importance of an individual's rank can be varied by changing the value of the ranking parameter q. This parameter makes the selection process less sensitive to the problem of premature convergence. Large values of q such as 0.4 result in more rapid convergence of the population while smaller values of q such as 0.05 tend to slow down convergence. The danger of a large q is that the evolution may converge prematurely on poor solutions. Conversely, a small q requires greater computation time.

(d) Crossover

At this stage a new population of survivors exists. The next step involves replacing the non-survivors through the genetic crossover function. A variety of crossover functions have been proposed. For example, in uniform crossover (Michalewicz, 1992), feature attribute values can be randomly copied to the child from one parent or the other until the child represents a complete design.

(d) Mutation

The final step is the mutation function. Mutation proceeds by randomly selecting a variable $a_{i,j,k}$ from a member i of the population and replacing it with a randomly generated value. The rate of mutation is one of the user selected parameters. This disruption of the population is done deliberately to introduce variety into the population. This variety helps the algorithm discover design alternatives that might not have otherwise been uncovered with crossover alone. However, mutation rates are normally kept very low, since high mutation rates reduce the algorithm to a random search technique.

The simplest form of mutation is uniform mutation in which the value of a variable is replaced with a randomly chosen value in that variable's domain. Alternatively, more sophisticated heuristics can be used to speed up convergence of the algorithm. For example, many numerical optimization problems have optimal solutions in which one or more variables have values at the upper or lower bounds of their domain. Hence one mutation operation involves changing a numerical variable's value to its upper or lower bound. This is referred to as boundary mutation (Michalewicz, 1992).

13.4.3 Research issue: genetic algorithms for concurrent engineering

The concurrent engineering area is one that is characterized by complexity and this complexity shows itself in determining viable and effective approaches to concurrent engineering systems. In particular, concurrent engineering problems are characterized by mixtures of algebraic, conditional, and database constraints with quantitative, symbolic and qualitative variables.

As regards GA applied to concurrent engineering, problems with constraints have received relatively little attention to date. Michalewicz and Attia (1994) has obtained good results in a system which addresses problems with nonlinear algebraic constraints. However, there has been no work to date on GA applied to problems with mixtures of algebraic, conditional and database constraints with quantitative, symbolic and qualitative variables. This, therefore, forms a significant research issue.

13.5 OVERVIEW OF THE CBGA CONCURRENT ENGINEERING SYSTEM

As indicated previously little work has been done on GA applied to constrained problems, certainly where there are, as with concurrent engineering problems, mixtures of algebraic, conditional, and database constraints with quantitative, symbolic and qualitative variables. The CBGA aims to address this deficiency and this section gives an overview of the CBGA system.

13.5.1 Architecture of the CBGA

The structure of the CBGA is shown in Fig. 13.1. The GA portion of the CBGA contains substantive extensions of the conventional GA to allow for mixtures of algebraic, conditional and database constraints with

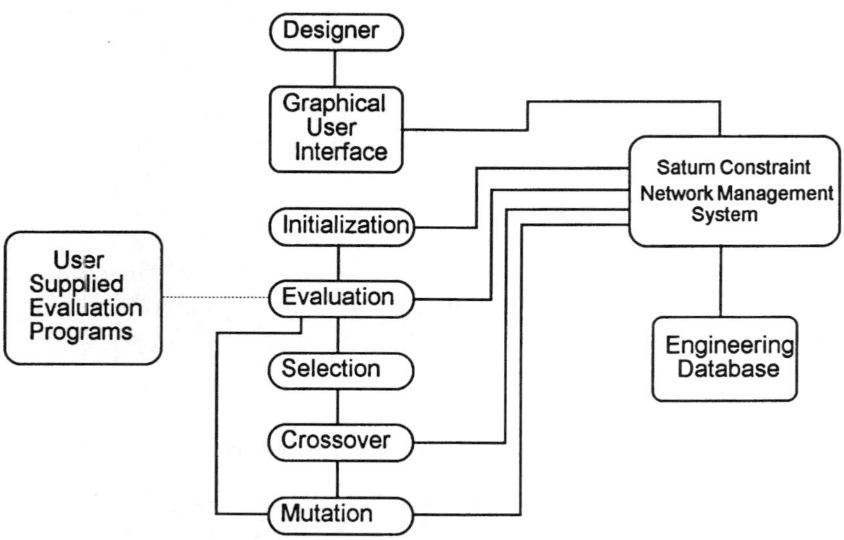

Fig. 13.1 Architecture of the CBGA concurrent engineering system.

quantitative, symbolic and qualitative variables. This GA portion is combined with a constraint modeling system.

13.5.2 Constrained genetic algorithms

As also indicated previously, the main deficiencies in conventional GAs occur in the *initialization, crossover* and *mutation* functions: the *evaluation* and *selection* functions in the existing work on GAs can be used in a relatively straightforward manner. An overview of the new **constrained** *initialization, crossover* and *mutation* algorithms in the CBGA follows. It should be noted that the following describes one instance of these algorithms; variations are possible and may perform better in certain circumstances.

(a) Constrained initialization

The *constrained initialization algorithm* in the CBGA strives for a balanced exploration of the search space and for an intelligent selection of the order in which variables are assigned values. This is done by the following steps.

1. Select variables from the constraints that subsume all other constraints (i.e. ignore redundant constraints).
2. Select the variables with the smallest domains.
3. Select the variables which appear in the largest number of constraints.

The population is also selected so that it is composed of viable members, i.e. members that satisfy all the constraints. This is done by propagating the effects of setting the variable values through the constraints using local constraint propagation and set propagation. This is achieved using the constraint network modeling system as shown in Fig. 13.1.

(b) Constrained crossover

The *constrained crossover algorithm* aims to intelligently select the order in which variables are copied from a parent to the offspring. This is done by the following steps.

1. Select variables from the constraints that subsume all other constraints.
2. Select the variables with the smallest domains.
3. Select the variables which appear in the largest number of constraints.

Having done this the algorithm uses appropriate operations depending on the type of variable. The variables can be real intervals, integer intervals and real, integer or symbolic sets.

Children are produced that are viable members, i.e. members that satisfy all the constraints. This is done, as with *initialization*, by propagating the effects of setting the variable values through the constraints using local constraint propagation and set propagation. This is done, as with initialization, using the constraint network modeling system as shown in Fig. 13.1.

(c) Constrained mutation

The *constrained mutation algorithm* operates by randomly retracting values of variables. When retracting a value for the variable being mutated, the algorithm also retracts the values of all variables which were propagated as a result of the mutated variable's original value. The algorithm then determines the mutated variable's domain and selects and assigns a new value accordingly. It should be noted that, as is the case with *crossover*, *mutation* can render a previously feasible solution infeasible. To overcome this problem, the algorithm propagates the consequences of the mutated value and assigns values to those resulting variables without values. This is done, as with initialization and crossover, using the constraint network modeling system as shown in Fig. 13.1. This process makes *mutation* more productive by helping to maintain feasible solutions.

(d) CBGA

The above *constrained initialization, crossover* and *mutation* algorithms form the basis of the CBGA. In essence, they substantially extend the conventional GA so that the CBGA can handle the constraints that are typically found in concurrent engineering problems. These include mixtures of algebraic, conditional and database constraints with quantitative, symbolic and qualitative variables.

13.5.3 Constraint network modeling system

The CBGA includes the Saturn constraint network modeling system (Fig. 13.1) and the portions of this section that describe the structure of Saturn are a synopsis of the description provided by Fohn *et al.* (1993).

The constraint-sheet interface serves as the environment's interface. Each cell in the sheet may contain either a domain, a constraint, a macro, a constraint-sheet formula, or a text string that can be used to annotate an application. A cell's domain defines the range of values that can be accommodated by the cell. Domains may be extensionally defined (explicitly enumerated) in any constraint-sheet cell or defined over a column in a database. When a domain, logic constraint, macro, or constraint-sheet is entered into the CBGA's Saturn module, the well-formed formulae compiler performs a number of checks to ensure that it is syntactically correct and that any references are correct. When a value is entered into a cell, the well-formed formulae compiler performs a syntax check to ensure that the value either exists in an extensionally defined CBGA domain or in a column of a referenced database table. All database interaction is performed via the relational database interface.

Constraint-variable values (i.e. constraint-sheet cell values), logic constraints, macros and constraint-sheet formulae are passed on to the constraint engine for inferencing during theorem proving. Logic constraints can accommodate a number of diverse atomic constraints such as numeric, inequality, Prolog and database constraints. Logic constraints implement constraint propagation and accommodate forward or top-down reasoning with CBGA applications. They are normalized into clause sets and are resolved using hyper-resolution with a singular, linear and ordered search strategy and a backtracking control mechanism. Propagation is achieved by asserting the cell-value hyper-resolvents in order to resolve the empty set, thereby proving satisfiable constraints. Satisfied logic constraints are those that have been determined to be *true*. Conversely, violated (or unsatisfied) logic constraints are those that have been determined to be *false*. If the constraint engine is incapable of determining the satisfiability or unsatisfiability of a logic constraint, the constraint remains *unknown*. The constraint engine also incorporates an equation evaluator that operates on constraint-sheet formulae and equations (atomic constraints comprising the equality predicate and mathematical terms). An equation rewriting system operates on equations to ensure that only functional expressions comprising ground terms are submitted to the equation evaluator. The inference engine treats macros and database constraints encountered during the proof process as goals.

The program evaluation performed by the inference engine provides a goal-directed or bottom-up reasoning feature. Macro clauses can be either standard Prolog clauses or database clauses. Standard Prolog clauses are evaluated by the Prolog interpreter, while database clauses are first transformed into equivalent SQL (Structured Query Language) sentences by a specialized unification algorithm. Theoretically this is achieved by taking a proof-theoretic view of the relational database management system (DBMS). These SQL sentences are then passed to the relational DBMS query processor, via the tightly-coupled *relational database interface*, for evaluation.

The assumption-based truth maintenance system (ATMS) is a non-monotonic theorem prover based on propositional logic that functions as an archive for all inferences performed and contradictions found by the constraint engine. The archived dependencies allow a user to view constraint violations and resolve conflicts by rolling back to an application's previous search state. Additionally, the ATMS can provide justifications for all variable values by presenting the solution path through which a cell's value was determined. It is this component of the CBGA, together with the visual truth-values denoting the state of the constraints in the constraint-sheet that provides feedback to a user when performing constraint satisfaction.

13.5.4 CBGA compared with other potential optimizing approaches

The CBGA has several distinct advantages when compared with other optimization techniques. The primary advantage is that the CBGA does not make restrictive assumptions regarding the nature or shape of the space being searched. This is very important since concurrent engineering problems typically involve real numbers, integers and symbolic values from manufacturing databases. These variables are combined to form objective functions which are heavily constrained by all the various concurrent engineering factors. The resulting decision space can be highly irregular (i.e. nonlinear with many discontinuities). Most calculus-based optimization techniques such as hill climbing require continuous and differentiable search spaces. Conversely, combinatorial optimization techniques such as dynamic programming require discrete variables. These assumptions regarding the nature of the problem render most optimization techniques inappropriate for concurrent engineering optimization. However, the GA makes no such assumptions regarding the problem and is therefore suitable for concurrent engineering optimization.

Another virtue of the GA is its ability to avoid being trapped in locally optimal solutions. The fact that the GA operates with a population of many different design points contributes to this robustness. The 'survival of the fittest' technique employed by the GA also ensures that the search for better solutions is relatively efficient. This balance between exploration and exploitation makes the GA ideal for large complex problems such as those found in concurrent engineering.

Finally, the fact that infeasible designs are permitted to evolve into feasible ones represents a significant departure from many traditional constrained optimization techniques which require feasible solutions as starting points. In heavily constrained problems like those in concurrent engineering, the computational cost associated with automatically finding completely feasible solutions can be prohibitive. It should also be noted that highly fit, yet slightly infeasible designs can contain large amounts of valuable information. Furthermore, the conflicting goals and constraints among the various concurrent engineering components can sometimes mean that no design alternatives exist which simultaneously satisfy all constraints. In these cases, it is necessary to search for designs which do the best job of making compromises.

13.6 A CBGA DESIGN SESSION

A prototype version of the CBGA has been implemented, and this section describes a simple yet illustrative design problem using the prototype CBGA. The problem involves the parametric design of a nutcracker subject to a variety of life-cycle constraints. A conceptual drawing of the nutcracker is shown in Fig. 13.2. The nutcracker assembly consists of two parts – a body and a thumbscrew. The body consists of a block with a slot to hold the nut while it is being cracked. The thumbscrew screws into the body through a threaded hole. As the thumbscrew is lowered, it exerts pressure on the nut until the nut cracks.

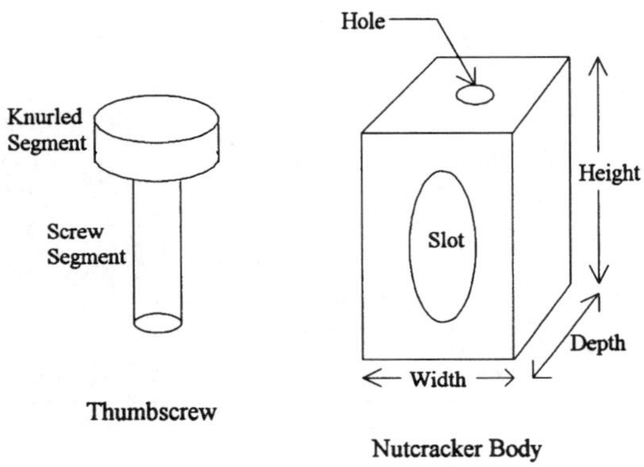

Fig. 13.2 Conceptual drawing of nutcracker.

We begin modeling the problem in the CBGA by identifying the features which characterize the nutcracker. For example, the nutcracker body can be described in terms of six surfaces, one slot and one threaded hole. The thumbscrew has one threaded segment and one knurled segment. Furthermore, each feature describing the nutcracker is specified in terms of a number of feature attributes. For example, the threaded hole in the body can be described by its diameter, length, position, thread designation etc. A portion of the nutcracker's feature-attribute description in the CBGA is shown in Fig. 13.3.

Note that each feature attribute corresponds to a separate CBGA variable. This means that for each attribute, the domain (i.e. all legal values) must be specified. The domain can be the set of real numbers or integers, strings, or can even come from tables in a manufacturing database. In practice, the database domain is extremely useful. For example, the nominal diameter in a thread designation might be specified as a real number while the number of threads per inch might be specified as an integer. However, a more practical approach used in our nutcracker example is to create a database table covering all standard thread designations for the tools in our machine shop. Note that through careful use of the database, a large number of constraints can be implicitly

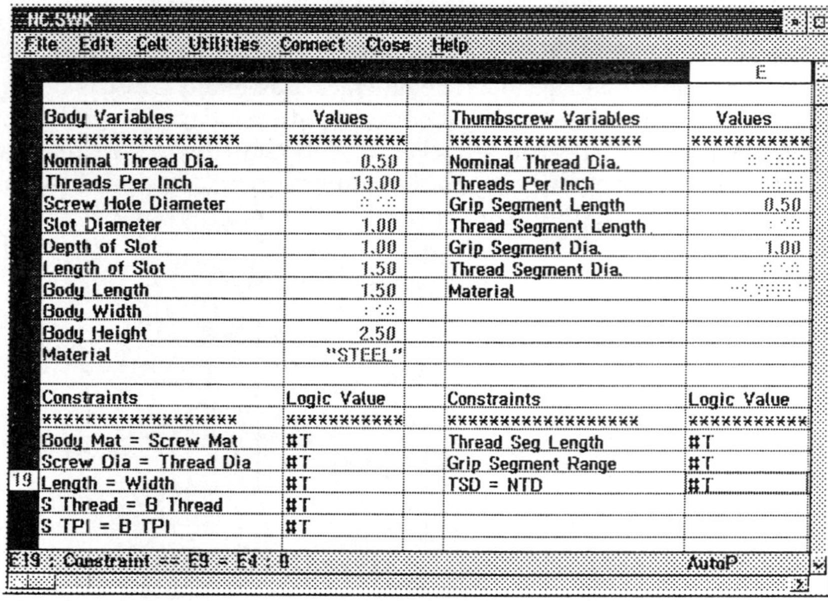

NC.SWK

File Edit Cell Utilities Connect Close Help

Body Variables	Values	Thumbscrew Variables	Values
******************	***********	******************	***********
Nominal Thread Dia.	0.50	Nominal Thread Dia.	[illegible]
Threads Per Inch	13.00	Threads Per Inch	[illegible]
Screw Hole Diameter	[illegible]	Grip Segment Length	0.50
Slot Diameter	1.00	Thread Segment Length	[illegible]
Depth of Slot	1.00	Grip Segment Dia.	1.00
Length of Slot	1.50	Thread Segment Dia.	[illegible]
Body Length	1.50	Material	[illegible]
Body Width	[illegible]		
Body Height	2.50		
Material	"STEEL"		
Constraints	Logic Value	Constraints	Logic Value
******************	***********	******************	***********
Body Mat = Screw Mat	#T	Thread Seg Length	#T
Screw Dia = Thread Dia	#T	Grip Segment Range	#T
Length = Width	#T	TSD = NTD	#T
S Thread = B Thread	#T		
S TPI = B TPI	#T		

E19 : Constraint == E9 = E4 : 0 AutoP

Fig. 13.3 Nutcracker feature attributes in a constraint sheet.

specified. In our thread designation example, we need not explicitly write constraints which specify the legal combinations of nominal diameter and threads per inch. They are already in the database.

While the database is useful for storing declarative facts which affect the design process, it is also necessary to explicitly write constraints in the CBGA. Life-cycle constraints on the nutcracker design can be logically separated into functional constraints, manufacturing constraints, ease of assembly constraints and reliability constraints. Some life-cycle constraints which a designer might wish to consider are shown in Table 13.1. This is a short list of some of the constraints which might be imposed on the nutcracker design. To calculate manufacturing costs, a series of Prolog macros were written which form a process planning system. Based on the feature attributes and the batch size, the macros select the appropriate manufacturing processes and estimate production costs accordingly. All other constraints are easily implemented using the CBGA's forward chaining logic constraints. For example, a constraint stating that the width and depth of the nutcracker body should be equal can be entered into a constraint sheet cell as '= Width = Depth' where the single '=' at the beginning of the formula tells the CBGA that the expression is a formula. Conversely, logic constraints are preceded by '= ='. Unlike a traditional

Table 13.1 Possible constraints on the nutcracker design

Constraint type	*Constraint name*	*Explanation*
Functional	Slot size	Is the slot wide enough, tall enough and deep enough to hold the desired type of nuts?
Functional	Screw torque	Is the screw grip large enough to generate sufficient torque for cracking the desired type of nuts?
Manufacturing	Body cost	What is the estimated manufacturing cost for the body?
Manufacturing	Thumbscrew cost	What is the estimated manufacturing cost for the thumbscrew?
Assembly	Thread size	Do the threaded hole and thumbscrew have matching thread designations?
Reliability	Thread strength	Based on material selection and thread designation, are the threads strong enough to withstand nut-cracking forces?
Reliability	Wall thickness	Are the body walls (determined by slot size) thick enough to withstand forces?

spreadsheet, this formula is propagated bi-directionally. For example, if Width is set to 1.5 inches (3.8 cm), Depth will automatically be assigned a value of 1.5 inches. Conversely, if Depth is assigned a value of 2.0 inches (5.1 cm), then Width is assigned a value of 2.0 inches.

Production system-style rules can be written as logic implications. For example, suppose that the material available for use include various types of hardwoods like oak or cherry in addition to aluminum and steel. We wish to add a constraint that says that the thumbscrew and body materials should not mix wood with metal. This can be stated in a constraint-sheet cell as follows:

= = Screw_Mat = Cherry OR Screw_Mat = Oak => Body_Mat
= Cherry OR Body_Mat = Oak

This constraint reads 'If the screw material is cherry or oak, then the body material should be either cherry or oak'. A similar constraint can be added for the case of aluminum or steel.

Once the problem and its constraints have been modeled in the CBGA, designers are free to manually assign values to any feature attributes. Any assignments made by designers are treated as fixed by the CBGA, and are not changed. Furthermore, designers can manually restrict the range of values over which feature attributes are defined. This allows designers to influence the direction of search taken by the CBGA by narrowing down the possibilities which can be explored.

All feature attribute values which are not fixed by the user are assigned values by the GA during the optimization phase. The graphical user interface of the CBGA's optimization module is shown in Fig. 13.4. This figure depicts intermediate results from a nutcracker optimization run. Throughout this optimization procedure, users are continually supplied with feedback. The line graph shown in Fig. 13.4 gives users the objective score of the best design alternative along with the population average. The population bar graph lets users know what the score of each member is for the current generation. This is used to visually indicate when the population has begun to converge on one or more good designs. The standard deviation of these scores is also recorded as vertical bars in the line graph so that users can more easily detect the overall trend in convergence.

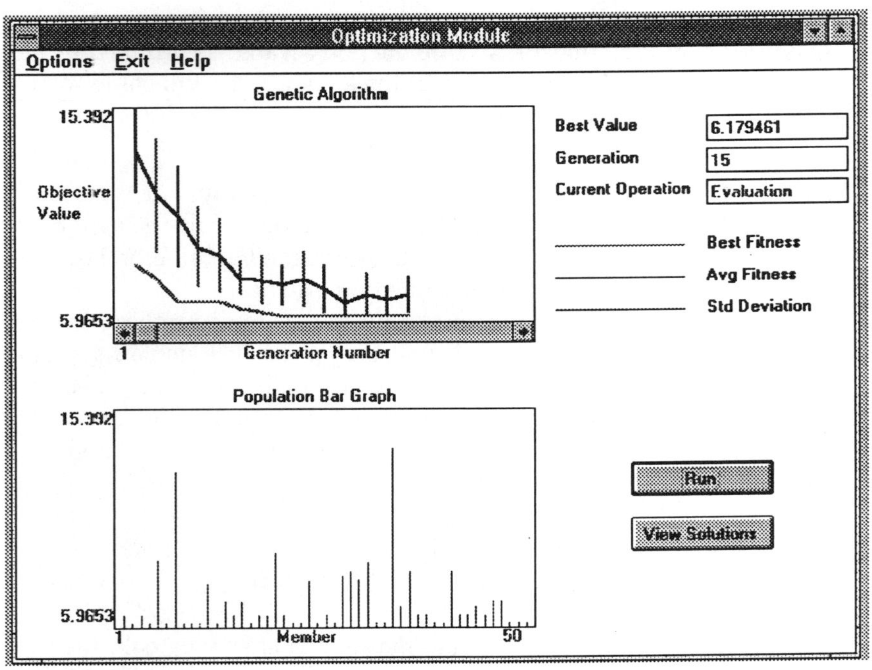

Fig. 13.4 Graphical user interface for CBGA's optimization module.

Once the design evolution process is complete, users have the option of viewing the various design alternatives discovered by the CBGA by clicking on the View Solutions button. Hence it is up to designers to make the final decision as to which alternative is truly optimal. In practice, this is important since the formulation of objective functions is not an exact science. Consequently, design alternatives with reasonably close scores should be evaluated by designers and should not be blindly chosen on the basis of the score alone.

In our nutcracker design problem, a CAD representation of the best alternative discovered by the CBGA is shown in Fig. 13.5. This particular alternative satisfied all life-cycle constraints placed on the design and also had the lowest estimated production cost as determined by the process planning rules.

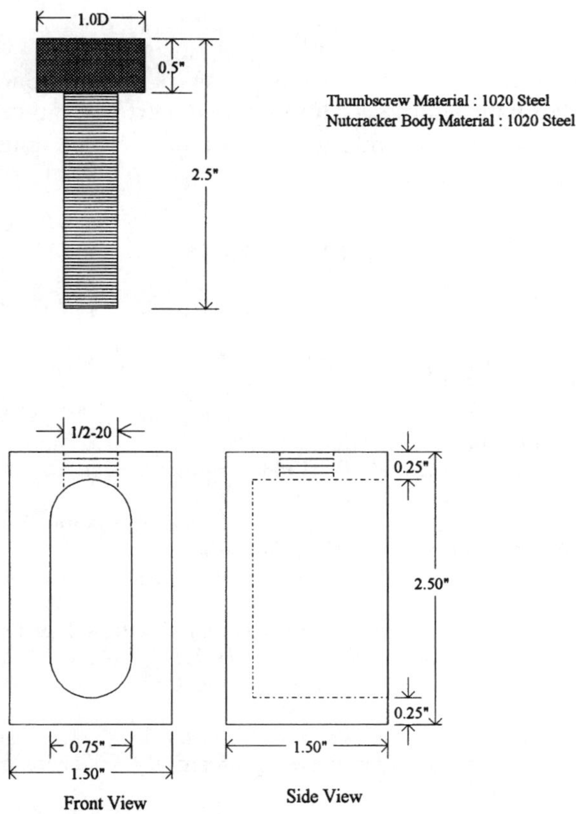

Fig. 13.5 Best found CBGA solution for nutcracker problem.

13.7 SUMMARY AND CONCLUSION

A key aspect of concurrent engineering is that of devising systems that determine improvements that can be made to a product. A new approach to design improvement is described in this paper. This approach is based on new algorithms that extend the conventional genetic algorithm to handle mixtures of algebraic, conditional and database constraints with quantitative, symbolic and qualitative variables. This approach we call the constraint-based genetic algorithm (CBGA).

The CBGA mimics the process of evolution by splicing together pieces of design alternatives in order to form new designs. During this process, CBGA ensures that no constraints are violated in order to maintain feasibility as much as is possible. By keeping only the best design alternatives, the algorithm employs a 'survival of the fittest designs' strategy.

ACKNOWLEDGEMENTS

The authors gratefully acknowledge the work done on Saturn by Arthur Greef, Steffen Fohn, Ramdas Murali and Robert E. Young.

This work was funded in part by National Science Foundation under Grant No. DDM-8914200.

REFERENCES

Andreasen, M.M., Khler, S. and Lund, T. (1983) *Design for Assembly*, IFS Publications, Bedford.

Bao, H.P. (1988) An expert system for SMT printed circuit board design for assembly. *Manufacturing Review*, **1**(4), 275–280.

Bedworth, D.D., Henderson, M.R. and Wolfe, P.M. (1991) *Computer-Integrated Design and Manufacturing*, McGraw-Hill.

Boothroyd, G. and Dewhurst, P. (1988) Product design for manufacture and assembly. *Manufacturing Engineering*, **100**(4), 42–46.

Borning, A. (1979) Thinglab – a constraint-oriented simulation laboratory. *XEROX PARC technical report SSL – 79–3*, Palo Alto, CA.

Bralla, J.G. (ed. in chief) (1986) *Handbook of Product Design for Manufacturing*, McGraw-Hill.

Chao, N.-H. (1985) The application of a knowledge based system to design for manufacture. *IEEE Document number CH2125–7/85/0000/0182.*

Chen, Y. and Young, R.E. (1988) PACIES (a part code identification expert system). *IIE Transactions*, **20**, 132–136.

Cohoon, J., Hedge, S., Martin, W. and Richards, D. (1991) Distributed genetic algorithms for the floorplan design problem. *IEEE Transactions on Computer-Aided Design*, **10**(4), 483–92.

Dechter, R. (1992) Constraint networks, in *Encyclopedia of Artificial Intelligence* (ed. S. Shapiro), John Wiley and Sons, New York, pp. 411–25.

DeJong, K. (1988) Learning with genetic algorithms: an overview, *Machine Learning*, **3**, 121–138.

Dieter, G.E. (1983) *Engineering Design: A Materials and Processing Approach.* McGraw-Hill, New York.

Evans, B. (1988) Simultaneous engineering. *Mechanical Engineering,* **110**(2), 38–39.

Eversheim, W. and Muller, W. (1984) Assembly oriented design in *International Trends in Manufacturing Technology: Programmable Assembly* (ed. W.B. Heginbotham), IFS Publications, Bedford.

Fohn, S., Greef, A., Young, R.E. and O'Grady, P. (1993) A constraint-system shell to support concurrent engineering approaches to design. *Artificial Intelligence in Engineering,* **9**(1), 1–17.

Fohn, S., Greef, A., Young, R.E. and O'Grady, P. (1994) A constraint based approach to AS/RS design, in *Proceedings IIE Research Conference,* May, Atlanta, IIE Press, pp. 473–478.

Greef, A., Fohn, S., Young, R.E. and O'Grady, P. (1994). Implementation of a logic-based support system for concurrent engineering. *Journal of Data and Knowledge Engineering,* **15**(1), 31–61.

Gross, M., Ervin, S., Anderson, J. and Fleisher, A. (1987) Designing with constraints, in *Computability of Design* (ed. E.K. Yehuda), John Wiley and Sons, New York, pp. 53–84.

Haeusler, J. (1981) Design for assembly – state of the art, in *Proceedings of the 2nd International Conference on Assembly Automation,* IFS (Conferences), Bedford.

Harfmann, A.C. (1987) The rationalizing of design, in *Computability of Design* (ed. E.K. Yehuda), John Wiley and Sons, New York, pp. 1–8.

Heintze, N., Michaylov, S. and Stuckey, P. (1987) CLP ($\mathfrak{R}$) and some electrical engineering problems, in *Logic Programming: Proceedings of Fourth International Conference,* (ed.) J.-L. Lassez, The MIT Press, Cambridge, MA, Vol. 2, pp. 675–703.

Heller, E.D. (1971) *Value Management: Value Engineering and Cost Reduction.* Addison-Wesley, Reading, MA.

Hollingurn, J. (1989) Expert system aids design for assembly. *Assembly Automation,* **9**(3), 132–136.

Husbands, P., Mill, F. and Warrington, S. (1991) Genetic algorithms, production plan optimization, and scheduling, in *Parallel Problem Solving From Nature, 1st Workshop,* (eds H.D. Schwefel and R. Manner) Springer-Verlag, New York, pp. 80–84.

Kim, Y. and O'Grady, P. (1996) A methodology for analyzing large scale concurrent engineering systems. *International Journal of Production Research* (to appear).

Kim, Y., O'Grady, P. and Young, R.E. (1991) A concurrent engineering system for turbine blade design. Department of Industrial Engineering Technical Report, North Carolina State University, Raleigh, North Carolina.

Kim, C., O'Grady, P. and Young, R.E. (1992a) TEST: a design for testability system for printed wiring boards. *Journal of Electronics Manufacturing,* **1**(2), 61–70.

Kim, J.-Y., Mittal, R., O'Grady, P. and Young, R.E. (1992b) Process selection for concurrent engineering in the domain of rotational parts. *Journal of Design and Manufacturing,* **2**, 199–209.

Kim, K., Cormier, D., O'Grady, P. and Young, R. (1995) A system for design and concurrent engineering under imprecision. *Journal of Intelligent Manufacturing,* **6**(1), 11–27.

Konopasek, M. and Jayaraman, S. (1984) *The TK! Solver Book,* Osborne/McGraw-Hill, Berkeley, CA.

Kroll, E., Lenz, E. and Wolfberg, J.R. (1988) A knowledge-based solution to the design for assembly problem. *Manufacturing Review,* **1**(2).

Lawton, G. (1992) Genetic algorithms for schedule optimization. *AI Expert*, **7**(5), 23–27.

Leler, W. (1988) *Constraint Programming Languages, Their Specification and Generation*, Addison-Wesley, Reading, MA.

Liau, J. and Young, R.E. (1992) A process planning and concurrent engineering system for printed circuit board assembly, in *The Second International Conference on Automation Technology*, Taipei, Taiwan.

Lu, C.-Y.S. (1986) Knowledge based expert systems: a new horizon for manufacturing automation. *Knowledge Based Expert Systems for Manufacturing*, **PED 24**, 11–24.

Mackworth, A. (1992) The logic of constraint satisfaction. *Artificial Intelligence*, **58** (1–3), 3–20.

Maddux, K.C. and Jain, S.C. (1986) CAE for the manufacturing engineer: the role of process simulation in concurrent engineering, in (eds A.A. Tseng, D.R. Durham and R. Komanduri) *Manufacturing Simulation and Processes* ASME, New York, 1–16.

Michalewicz, Z. (1992) *Genetic Algorithms + Data Structures = Evolution Programs*, Springer-Verlag, Berlin, pp. 83–96.

Michalewicz, Z. and Attia, N. (1994) Evolutionary optimization of constrained problems, in *Proceedings of the 3rd Annual Conference on Evolutionary Programming* (eds A.V. Sebald and L.J. Fogel); River Edge, NJ, World Scientific Publishing, pp. 98–108.

Nelson, G. (1985) JUNO, A constraint-based graphics system, *SIG-GRAPH Computer Graphics*, **19**(3), 235–43.

Oakley, M. (1984) *Managing Product Design*. John Wiley and Sons, New York.

Oh, J.S., O'Grady, P. and Young, R.E. (1995) An artificial intelligence constraint network approach to design for assembly, *IIE Transactions*, **27**(1), 72–80.

O'Grady P. and Oh, J.S. (1990) Arthur – an intelligent system for design for assembly, concurrent engineering of mechanical systems, in *Proceedings of the Second Annual Symposium on Mechanical System Design in a Concurrent Engineering Environment*, Iowa City, Iowa, pp. 293–306.

O'Grady, P. and Oh, J.S. (1991) A review of approaches to design for assembly. *Concurrent Engineering*, **1**, 5–11.

O'Grady, P. and Young R. (1991) Issues in concurrent engineering systems, *Journal of Design and Manufacturing*, **1**, 27–34.

O'Grady, P., Kim, Y. and Young, R.E. (1994) A hierarchical approach to concurrent engineering systems. *International Journal of Computer Integrated Manufacturing*, **7**, (3) 152–62.

O'Grady, P., Young, R. and Smith, L. (1991) An advice system for concurrent engineering, *International Journal of Computer Integrated Manufacturing*, **4** (2) 63–70.

Redford, A.H. (1986) Software aid to design for assembly. *Assembly Automation*, May.

Redford, A.H., Swift, K.G. and Howie, R. (1981) Product design for automatic assembly, in *Proceedings Second International Conference on Assembly Automation*, IFS (Conferences), Bedford.

Rosen, D., Riley, D. and Erdman, A. (1987) A general knowledge-based system shell, with application to dwell mechanism design in *Proceedings of the International Computers in Engineering Conference*, Vol. 1, New York, August, ASME, New York, pp. 29–36.

Sackett, P.J. and Holbrook, A.E. (1988) DFA as a primary process decreases design deficiencies. *Assembly Automation*, 8(3), 137–40.

Sevenler, K., Reghupathi, P.S., Altan, T. and Miller, R.A. (1986) Knowledge-based

system approach to forming sequence design for cold forging. *Knowledge Based Expert Systems for Manufacturing*, **PED-24** 299–310.
Stefik, M. (1981a), Planning with constraints (MOLGEN: Part 1). *Artificial Intelligence*, **16**, 111–40.
Stefik, M. (1981b) Planning and meta-planning (MOLGEN: Part 2), *Artificial Intelligence*, **16**, 141–70.
Stein, J. and Strasser, F. (1986) Metal stampings, McGraw-Hill, (Ed. in chief) J.G. Bralla in *Handbook of Product Design for Manufacturing*, pp. 3–13 to 3–36.
Sussman, G. and Steele, G. (1980) CONSTRAINTS – A language for expressing almost-hierarchical descriptions. *Artificial Intelligence*, **14**, 1–39.
Sutherland, I. (1963) SKETCHPAD: a man-machine graphical communication system, in *IFIPS Proceedings of the Spring Joint Computer Conference.*
Swift, K. (1987) *Knowledge Based Design for Manufacture*, Kogan Page, London.
Tam, K.Y. (1992) Genetic algorithms, function optimization, and facility layout design, *European Journal of Operational Research*, **63**(2), 322–46.
Trucks, H.E. (1987) *Designing for Economical Production*, Society of Manufacturing Engineers, Dearborn, MI.
Van Hentenryck, P. (1989) *Constraint Satisfaction in Logic Programming*, The MIT Press, Cambridge, MA.
Van Wyk, C. (1981) *IDEAL User's Manual*, Bell Labs, Computer Science Technical Report 103.
Vancza, J. and Markus, A. (1991) Genetic algorithms in process planning, *Computers In Industry*, **17** (2), 181–94.
Whitley, D. (1989) The GENITOR Algorithm and selection pressure: why rank-based allocation of reproductive trials is best, in *Proceedings of the Third International Conference On Genetic Algorithms* (ed. J.D. Schaffer), Morgan Kaufman, San Mateo, CA, pp. 116–21.
Wu, P. (1988) Design for testability, in *Proceedings, National Conference of the American Association for Artificial Intelligence*, pp. 358–63.
Young, R.E., Greef, A. and O'Grady, P. (1992) SPARK: an artificial intelligence constraint network system for concurrent engineering. *International Journal of Production Research*, **30**(7), pp. 1715–35.
Zorowski, C.F. (1986) PDM – a product assemblability merit analysis tool. Presented at the Design Engineering Technical Conference, Columbus, Ohio, October 5–8.

CHAPTER FOURTEEN

Computer-integrated manufacturing: a complex information system

Reuven R. Levary

14.1 INTRODUCTION

To be successful in the competitive global environment, manufacturers must be able to respond quickly to changes in the manufacturing environment. Automation technology makes it possible for manufacturers to have this essential capability. A conceptualized information system that integrates the programmable design of a product and its manufacture with both planning activities and the manufacturing process will be described in this chapter. The information system integrates the programmable automation technologies used in computer-integrated manufacturing with engineering and business databases. The computer-integrated manufacturing concept is expanded to the concept of a computer-integrated economy. The technological difficulties that arise when implementing a computer-integrated economy will be considered.

14.2 OVERVIEW OF PROGRAMMABLE TECHNOLOGIES USED IN COMPUTER-INTEGRATED MANUFACTURING

A manufacturer must respond quickly to changes in the manufacturing environment in order to be successful in today's competitive global environment. Automation technology is essential to the achievement of such a quick response time. A manufacturing system is considered automated if it is to some degree self-acting, self-regulating and self-dependent. The less dependent a manufacturing system is on a human operator, the higher its degree of automation.

Hardware that allows for programmable automation is one component that contributes to a manufacturer's ability to respond quickly to changes in the manufacturing environment. **Programmable automation** or **flexible automation** is the technology that makes it possible for the processing sequence and type of operations to be easily altered by means of software. Hardware involved in the manufacturing process of a product that allows for such automation includes the following.

1. Computerized numerical control (CNC) machine tools: these are used in manufacturing the individual parts of a product. The technology of CNC uses software to automatically control the operations of machines. This vital technology is the essence of automation. Some technical details of CNC systems are described in the *Standard Handbook of Industrial Automation* edited by Considine (1986).
2. Specialized robots: these are used in loading and unloading machine tools, inspecting parts, assembling products or performing manufacturing tasks like painting and welding. Robotics is well presented by Koren (1985).
3. Material transfer systems, such as networks of automated guided vehicles (AGVs): these are used in moving raw materials, workpieces, partially finished and finished parts, and tools through the manufacturing facility. An automated material transfer system (AMTS) also moves finished parts into an automated storage and retrieval system (ASRS) and moves various materials, components and tools out. Automated material transfer systems are described by Groover (1987) and Mitchell (1991).
4. Storage and retrieval systems: these are used for organizing valuable storage space. Such systems provide high-rise, high-density storage. An overview of automated storage and retrieval systems is given by Groover (1987).

The flow of materials into, out of and within an automated manufacturing facility is outlined. Raw materials delivered by suppliers are transferred by an AMTS into an ASRS. When needed, the AMTS moves all necessary materials from the ASRS to the work stations where individual parts are manufactured. Finished parts are transferred by the AMTS to temporary storage in the ASRS. When the manufacturer is ready to assemble the product, the AMTS moves the parts from the ASRS to the assembly line. Finished products are moved by the AMTS to temporary storage in the ASRS until the manufacturer is ready for distribution. Ultimately, the product is distributed to the consumers via a distribution network. Depending upon the needs of the manufacturer, storage may be bypassed on occasion. For example, raw materials may move directly to work stations and/or finished products may move directly into the distribution network.

The programmable design of a product and its manufacture is another important component contributing to a manufacturer's ability to respond quickly to changes in the manufacturing environment. This component encompasses the design of a product as well as design of the manufacturing facility and the manufacturing process. **Programmable design** or **flexible design** is the technology that makes it possible for design characteristics of a part, product, system or process to be easily altered by means of software. The essence of this technology is an interactive computer model that mimics the part, product, system or process that is to be designed. Once such a model is available, design engineers can develop alternative designs merely by interacting with the computer model via the computer keyboard and/or mouse. Design engineers can also use the computer model to analyze and evaluate the performance of each alternative design.

When an interactive computer graphics system is used for designing, the system is called computer-aided design (CAD). In this case, an interactive computer-animated model mimics the part, product or system. CAD technology can significantly reduce the time involved in the product design process. Additionally, it can provide a detailed computer analysis of the expected product performance. CAD technology can even be used to effectively design a manufacturing facility. A computerized interactive animated model allows rapid coordination of the movements of those machines, equipment, robots and materials involved in the manufacturing process of a product. In this way, collisions can be avoided through computer planning. CAD technology is described by Mitchell (1991).

The design of a product's manufacturing process is done in two phases. In Phase I, manufacturing engineers first determine the machines, tools, robots and operations needed for manufacturing the parts and assembling the product. Once that step is complete, engineers and/or operations researchers design the sequence of operations, material movements and assembly steps needed in manufacturing the product. The sequence is designed to optimize one or more measures of effectiveness of the manufacturing process. For example, the sequence might be designed to minimize machine idle time, total manufacturing time or total production cost. Computer-aided process planning (CAPP) is the programmable design technology that was developed to improve the effectiveness of designing this phase of the manufacturing process. It helps to reduce the time needed for designing such processes. CAPP technology is described by Groover (1987).

In designing the manufacturing process, the final step of Phase I is the design of on-line inspection activities. This includes designing the ways in which the results of the inspection activities are used to adjust the manufacturing process on-line. Computer-aided quality control (CAQC) is the automation technology used in automating the inspection activities and the integration of the inspection activities with the manufacturing process.

Phase II is design of the software used for controlling the manufacturing process on-line. This includes software that:

1. controls the operations of each machine, robot and equipment used in the manufacture of a product (e.g. part programs for CNC machine tools (Kral, 1986) and task programs for robots (Koren, 1985));
2. integrates the operations of all computer-controlled machines, robots and equipment used in the manufacture of a product.

Computer-aided manufacturing (CAM) is defined as the use of computers to improve both the effectiveness and efficiency of manufacturing activities. CAM can be divided into on-line and off-line applications. On-line applications include use of software to control the operations of each machine, robot and equipment used in the manufacturing process (e.g. part programs, task programs), and use of software to integrate the operations of all computer-controlled hardware involved in a product's manufacturer. Off-line applications include preparation of the above-mentioned software, and planning of activities that affect the manufacture of a product (e.g. production, inventory, material and cash-flow planning). CAM technology is presented by Groover (1987) and by Mitchell (1991).

14.3 A CONCEPTUALIZED INFORMATION SYSTEM FOR COMPUTER- INTEGRATED MANUFACTURING

The integration of the operations of all computer-controlled hardware involved in a product's manufacture with the programmable design of the product and its manufacture forms a complex information system. The essence of such an information system is its database. The database contains engineering characteristics of the product (e.g. geometry), of the hardware used in the manufacture of the product and of the manufacturing process. The design process of a product and its manufacture is directly linked with this database. Data can be used, added, deleted or modified as needed during the design process. Consider as an example the preparation of a part program for a CNC machine tool using a CAM system. Part programs specify the geometry for the part, the machining parameters, the sequence of operations to be performed on the part and the tool position at each instant in time during the manufacturing process of the part. When CAM software is integrated with CAD software, a single database can be shared. For example, the database that contains the geometric characteristics of a part can be shared. The geometric characteristics of the part are established using the CAD system. The database is periodically modified by the CAD system during the design process until the time that the final design is established. The CAM software uses the data that characterize the part geometry as input to the programming software that specializes in programming part programs (e.g. APT (Kral, 1986)).

A conceptualized information system (IS) that integrates the programmable design of a product and its manufacture is illustrated in Fig. 14.1. The database of this IS contains engineering data related to the product and its manufacture. As shown in the figure, the database comprises three distinct components. The first component includes data regarding product characteristics; the second component includes data regarding the manufacturing facility characteristics; and the third includes data regarding the characteristics of the manufacturing process. The change of data in one component may lead to changes in the data of the other component. Thus, the data components are distinct yet interdependent. For example, changes in product characteristics may necessitate changes in the characteristics of the manufacturing process. Changes in the manufacturing process may lead to changes in the layout of the computer-controlled hardware used in the manufacturing process (e.g. CNC machine tools, robots, AMTS) and, in turn, lead to changes in the data characterizing the manufacturing facility.

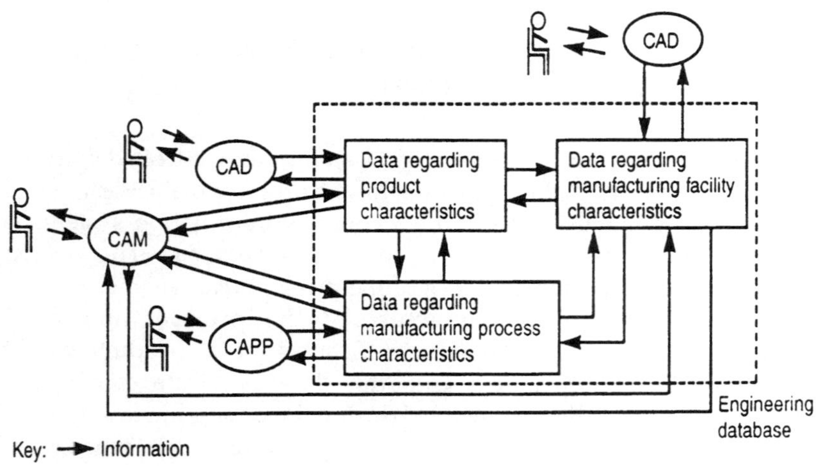

Fig. 14.1 A conceptualized information system integrating the programmable design of a product and its manufacture.

Engineers interact with the engineering database through a CAD system as they design or modify the product and/or manufacturing facility. Similarly, engineers and/or operations researchers interact with the engineering database through a CAPP system as they design or modify the manufacturing process. Using a CAM system, both engineers and/or programmers interact with the engineering database

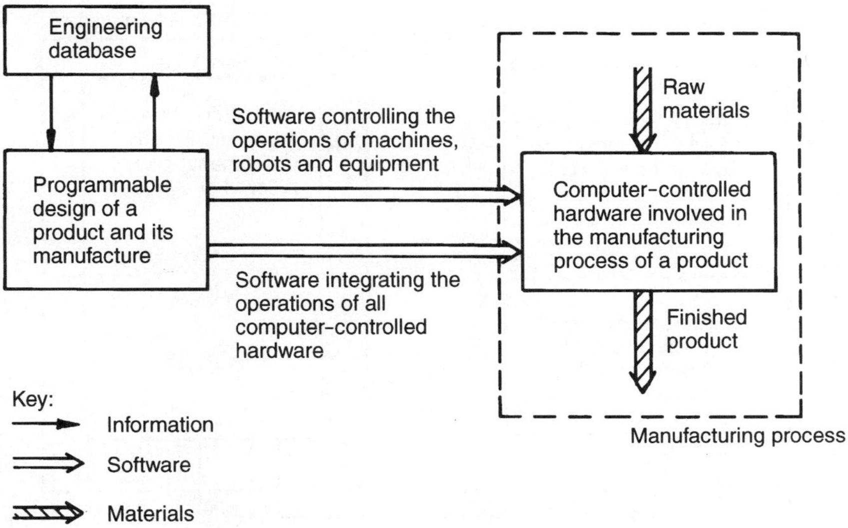

Fig. 14.2 Schematic representation of the integration of the manufacturing process with the programmable design of a product and its manufacture.

as they develop and update the software that controls the manufacturing process.

The integration of the manufacturing process with the programmable design of a product and its manufacture is schematically represented in Fig. 14.2. As can be seen from the figure, the outcome of the programmable design of a product and its manufacture is software that controls the manufacturing process.

The integration of the manufacturing process with information regarding non-functioning equipment, as well as information resulting from on-line automation inspection activities, allows automatic adjustment of a manufacturing process. The integration is schematically represented in Fig. 14.3. Non-functioning equipment results from equipment failure and/or need for maintenance. Automatic adjustment of a manufacturing process necessitates alteration of the software that controls the manufacturing process. Because automatic alteration of this software is a complex task, expert system technology may be needed.

The sharing of a single engineering database when designing and/or modifying a product, manufacturing facility or the manufacturing process contributes to a significant reduction in human error. Data that must be transferred with the help of humans, to the contrary, inevitably contribute to an increase in human error.

The integration of an automated manufacturing process with the

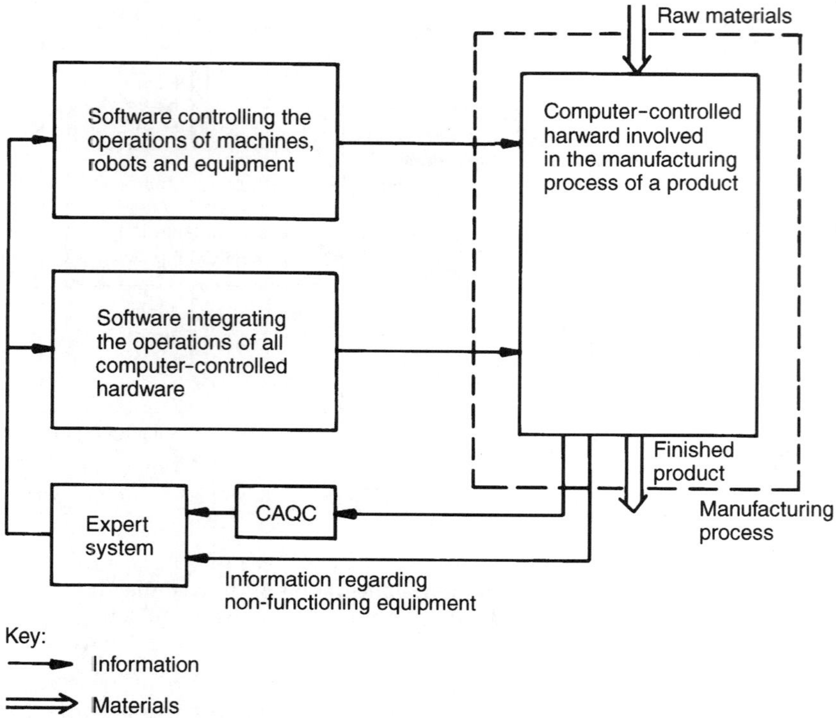

Fig. 14.3 Schematic representation of the integration of the manufacturing process with information regarding on-line automatic inspection activities and non-functioning equipment.

programmable design of a product and its manufacture contributes to a substantial shortening of the time between product conception and product completion. This enables manufacturers to respond more rapidly to changes in the manufacturing environment. Further reduction in response time can be achieved by also integrating planning activities with the design of a product and its manufacture. Planning activities include planning of:

1. production levels;
2. inventory levels of raw materials, partially finished parts, finished parts and finished products;
3. orders of raw materials and components' parts;
4. cash flows;
5. promotions.

A conceptualized IS that integrates the programmable design of a product and its manufacture with both the planning activities and the

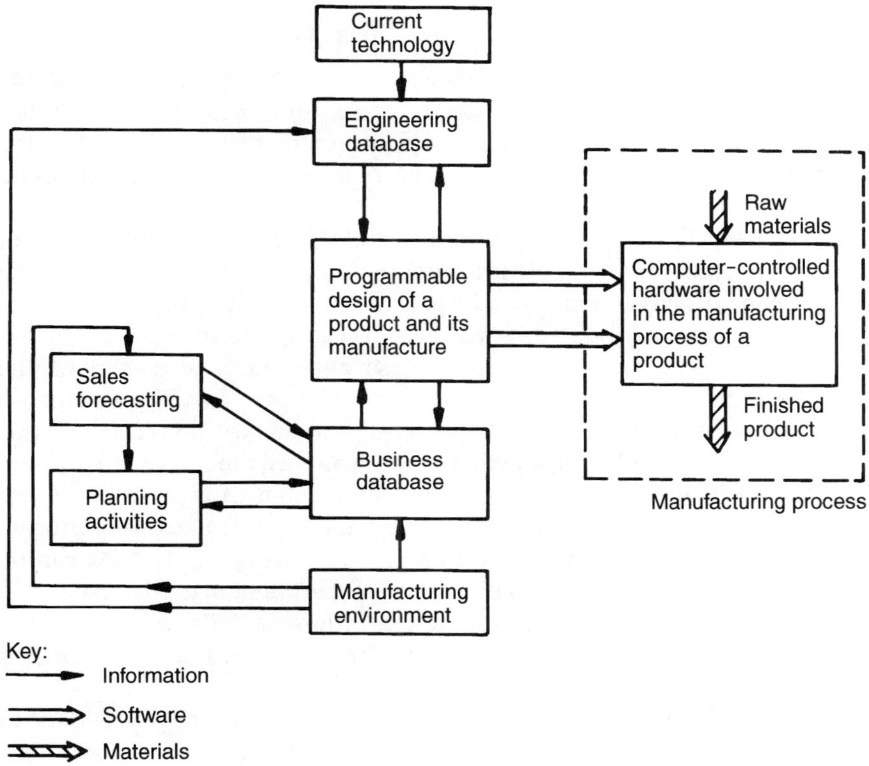

Fig. 14.4 A conceptualized information system integrating the programmable design of a product and its manufacture with both planning activities and the manufacturing process.

manufacturing process is illustrated in Fig. 14.4. The IS has two distinct databases, an engineering database and a business database. As mentioned earlier, the engineering database includes data regarding product characteristics, manufacturing facility characteristics and characteristics of the manufacturing process. The engineering database also contains data regarding current technology as well as pertinent data related to the current manufacturing environment such as trends in style of product design. The business database includes data regarding:

1. production capacity;
2. inventory levels;
3. supply of raw materials and partially finished parts;
4. distribution network;
5. sales and marketing;
6. cash flow;

7. accounting system;
8. competition.

The business database is periodically updated. Changes in the manufacturing environment, sales forecasts and/or planning activities make such an update imperative. Computer models used in planning such things as production levels, inventory levels and the level and timing of orders are described in Nahmias (1989).

The integration of planning activities with the programmable design of a product and its manufacture allows automatic adjustment of both production volume and production mix whenever either volume or production mix is updated. It was mentioned earlier that an automatic adjustment of a manufacturing process necessitates alteration of the software that controls the manufacturing process. The IS, represented schematically in Fig. 14.4, allows such automatic alteration of this software each time the production volume and mix are updated.

An IS that integrates the programmable design of a product and its manufacture with both planning activities and the manufacturing process is referred to as computer-integrated manufacturing (CIM). CIM can be defined as the automated integration of all pertinent activities related to both a product and its manufacture. More details regarding the characterization of CIM are given by Groover (1987) and by Mitchell (1991).

The IS illustrated schematically in Fig. 14.4 can be expanded to incorporate automatic coordination of manufacturing activities with distributors and suppliers. This is illustrated in Fig. 14.5. Automatic transfer of information regarding product distribution and sales improves the effectiveness of a CIM system. Automatic exchange of information with suppliers regarding orders and delivery schedules improves the effectiveness of a CIM system as well.

The importance of automatic transfer of information regarding product distribution and sales is apparent in the case of Benetton. The case was described first by Heskett and Signorelli (1989) and later by Clark (1990). Benetton is an Italian fashion company that uses an advanced IS to link retail stores, warehouses, flexible factories and suppliers. As soon as changes in consumer taste (e.g. colors, fad items) are recognized, production schedules are rapidly adjusted to accommodate the new trends in the market.

14.4 TOWARD THE CHALLENGE OF A COMPUTER-INTEGRATED ECONOMY

Computer-integrated manufacturing was previously defined as the automated integration of all activities related to both a product and its

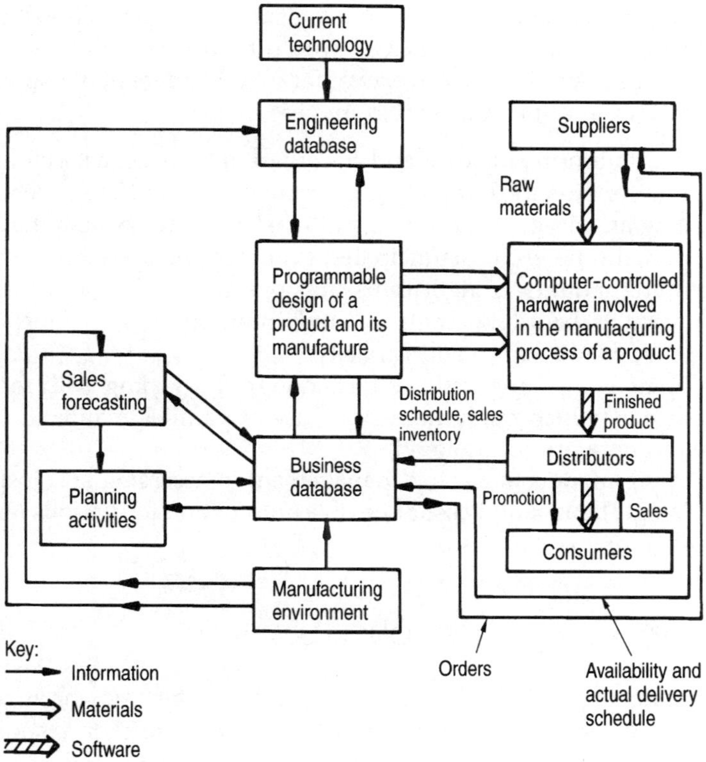

Fig. 14.5 Schematic representation of an information system that links the manufacturing process, the programmable design of a product and its manufacture, the suppliers, the distributors and the consumers.

manufacture. To enhance the effectiveness of CIM technology, communication with retail stores, warehouses and suppliers must also be automated and integrated with manufacturing activities. To make automated integration most effective, however, the automated communication network must extend even further. Suppliers, for example, must automate their own operations if they are to exhaust the potential of the automated linkage between themselves and the manufacturer. Thus, one successful automated company creates the need for another, and eventually an entire computer-integrated mini-economy is established. Components of such an economy would include manufacturers, suppliers, warehouses, retail stores, transportation companies, pertinent service industries and consumers having interactive TV.

To keep pace with unpredictable occurrences in the marketplace as well as with fast changes in competition, consumer taste and the economic environment, a computer-integrated economy (CIECON) would need a high degree of flexibility. The technological difficulties that would be encountered in designing a CIECON would be conceptually identical to those found in designing a CIM, except for a higher degree of complexity. The technological difficulties would involve:

- designing uniform software and hardware interfaces, as well as data exchange formats for integrating equipment from different vendors;
- designing an adaptive and intelligent master control system that could cope in real time with unanticipated events in the system;
- developing adaptive algorithms for on-line, real-time control of sequencing, scheduling, material and information flow, inspection, preventive maintenance and capacity;
- developing computing systems fast enough for dealing with real-time calculations (which might necessitate use of parallel architecture and/or neural network technology);
- developing models to aid top management in assessing and justifying the strategic benefit and cost of flexible automation for various levels of system complexity and size.

14.5 CONCLUSIONS

Incorporating programmable automation technology into the manufacturing process enables a manufacturer to have a quicker response to changes in the manufacturing environment. The shorter a manufacturer's response time, the more competitive the manufacturer. Various types of programmable automation hardware were described in this chapter. Also described were several technologies for programmable design of products and their manufacturing process.

A complex information system was described that integrates the operations of all computer-controlled hardware involved in a product's manufacture with the programmable design of the product and its manufacturing process. Such an information system is referred to as computer-integrated manufacturing. The role of engineering and business databases in the design and usage of such information systems was also described. The concept of computer-integrated manufacturing was expanded to the concept of a computer-integrated economy. Various technological difficulties related to the design and implementation of a computer-integrated economy were identified.

REFERENCES

Clark, K.B. (1990) What strategy can do for technology, *Harvard Business Review*, Nov.–Dec., 94–98.

Considine, D.M. (ed.) (1986) *Standard Handbook of Industrial Automation*, Chapman & Hall, New York.
Groover, M.P. (1987) *Automation Production Systems, and Computer Integrated Manufacturing*, Prentice-Hall, Englewood Cliffs, NJ.
Heskett, J.L. and Signorelli, S. (1989) *Benetton (A), 9–685–014*, Harvard Business School Publishing Division, Boston.
Koren, Y. (1985) *Robotics for Engineers*, McGraw-Hill, New York.
Kral, I.W. (1986) *Numerical Contol Programming in APT*, Prentice-Hall, Englewood Cliffs, NJ.
Mitchell, F.H., Jr (1991) *CIM Systems: An Introduction to Computer-Integrated Manufacturing*, Prentice-Hall, Englewood Cliffs, N.J.
Nahmias, S. (1989) *Production and Operations Analysis*, Irwin, Homewood, IL.

Index

Page numbers in *italics* refer to tables and page numbers in **bold** refer to figures